ARTIFICIAL INTELLIGENCE FOUNDATIONS

BCS, THE CHARTERED INSTITUTE FOR IT

BCS, The Chartered Institute for IT, is committed to making IT good for society. We use the power of our network to bring about positive, tangible change. We champion the global IT profession and the interests of individuals, engaged in that profession, for the benefit of all.

Exchanging IT expertise and knowledge
The Institute fosters links between experts from industry, academia and business to promote new thinking, education and knowledge sharing.

Supporting practitioners
Through continuing professional development and a series of respected IT qualifications, the Institute seeks to promote professional practice tuned to the demands of business. It provides practical support and information services to its members and volunteer communities around the world.

Setting standards and frameworks
The Institute collaborates with government, industry and relevant bodies to establish good working practices, codes of conduct, skills frameworks and common standards. It also offers a range of consultancy services to employers to help them adopt best practice.

Become a member
Over 70,000 people including students, teachers, professionals and practitioners enjoy the benefits of BCS membership. These include access to an international community, invitations to a roster of local and national events, career development tools and a quarterly thought-leadership magazine. Visit www.bcs.org/membership to find out more.

Further information
BCS, The Chartered Institute for IT,
3 Newbridge Square,
Swindon, SN1 1BY, United Kingdom.
T +44 (0) 1793 417 417
(Monday to Friday, 09:00 to 17:00 UK time)
www.bcs.org/contact
http://shop.bcs.org/

ARTIFICIAL INTELLIGENCE FOUNDATIONS
Learning from experience

Andrew Lowe and Steve Lawless

© the authors 2021

The right of Andrew Lowe and Steve Lawless to be identified as authors of this work has been asserted by them in accordance with sections 77 and 78 of the Copyright, Designs and Patents Act 1988.

All rights reserved. Apart from any fair dealing for the purposes of research or private study, or criticism or review, as permitted by the Copyright Designs and Patents Act 1988, no part of this publication may be reproduced, stored or transmitted in any form or by any means, except with the prior permission in writing of the publisher, or in the case of reprographic reproduction, in accordance with the terms of the licences issued by the Copyright Licensing Agency. Enquiries for permission to reproduce material outside those terms should be directed to the publisher.

All trade marks, registered names etc. acknowledged in this publication are the property of their respective owners. BCS and the BCS logo are the registered trade marks of the British Computer Society charity number 292786 (BCS).

Published by BCS Learning and Development Ltd, a wholly owned subsidiary of BCS, The Chartered Institute for IT, 3 Newbridge Square, Swindon, SN1 1BY, UK.
www.bcs.org

Paperback ISBN: 978-1-78017-5287
PDF ISBN: 978-1-78017-5294
ePUB ISBN: 978-1-78017-5300
Kindle ISBN: 978-1-78017-5317

British Cataloguing in Publication Data.
A CIP catalogue record for this book is available at the British Library.

Disclaimer:
The views expressed in this book are of the authors and do not necessarily reflect the views of the Institute or BCS Learning and Development Ltd except where explicitly stated as such. Although every care has been taken by the authors and BCS Learning and Development Ltd in the preparation of the publication, no warranty is given by the authors or BCS Learning and Development Ltd as publisher as to the accuracy or completeness of the information contained within it and neither the authors nor BCS Learning and Development Ltd shall be responsible or liable for any loss or damage whatsoever arising by virtue of such information or any instructions or advice contained within this publication or by any of the aforementioned.

All URLs were correct at the time of publication.

Publisher's acknowledgements
Reviewers: Sarah Burnett, Christine Burt
Publisher: Ian Borthwick
Commissioning editor: Rebecca Youé
Production manager: Florence Leroy
Project manager: Sunrise Setting Ltd
Copy-editor: The Business Blend Ltd
Proofreader: Barbara Eastman
Indexer: Matthew Gale
Cover design: Alex Wright
Cover image: istock/Lovegtrv6
Typeset by Lapiz Digital Services, Chennai, India

CONTENTS

Figures and table	vii
Authors	ix
Acknowledgements	x
Abbreviations	xi
Glossary	xiii
Useful websites	xix
Preface	xxi

1. INTRODUCTION: ETHICAL AND SUSTAINABLE HUMAN AND ARTIFICIAL INTELLIGENCE — 1
  1.1 The general definition of human intelligence — 1
  1.2 Definition of artificial intelligence — 3
  1.3 A brief history of artificial intelligence — 4
  1.4 Sustainable AI — 23
  1.5 Machine learning – a significant contribution to the growth of artificial intelligence — 24
  1.6 Summary — 25

2. ARTIFICIAL INTELLIGENCE AND ROBOTICS — 26
  2.1 Understanding the AI intelligent agent — 26
  2.2 What is a robot? — 31
  2.3 What is an intelligent robot? — 36
  2.4 Summary — 36

3. APPLYING THE BENEFITS OF AI AND IDENTIFYING CHALLENGES AND RISKS — 37
  3.1 Sustainability – how our values will change humans, society and organisations — 37
  3.2 The benefits of artificial intelligence — 39
  3.3 The challenges of artificial intelligence — 43
  3.4 Understanding of the risks AI brings to projects — 44
  3.5 Opportunities for AI — 46
  3.6 Funding AI projects – the NASA Technology Readiness Levels — 47
  3.7 Summary — 48

4. STARTING AI: HOW TO BUILD A MACHINE LEARNING TOOLBOX — 49
  4.1 How we learn from data — 49
  4.2 Neural networks — 80
  4.3 Agents' functionality — 85

	4.4 Using the cloud – cheap high-performance computing for AI	86
	4.5 Summary	87
5.	**ALGORITHMS**	**88**
	5.1 What is an algorithm?	89
	5.2 What is special about AI/ML algorithms?	90
	5.3 What are self-learning algorithms?	90
	5.4 What are the algorithms used in machine learning and artificial intelligence?	90
	5.5 What is a deep learning algorithm?	93
	5.6 The ethical use of algorithms	94
	5.7 Summary	95
6.	**THE MANAGEMENT, ROLES AND RESPONSIBILITIES OF HUMANS AND MACHINES**	**96**
	6.1 Artificial intelligence will drive humans and machines to work together	96
	6.2 Future directions of humans and machines working together	99
	6.3 'Learning from experience' Agile approach to projects	99
	6.4 Summary	101
7.	**AI IN USE IN INDUSTRY: REIMAGINING EVERYTHING IN THE FOURTH INDUSTRIAL REVOLUTION**	**102**
	7.1 AI already in use	102
	7.2 Summary	118
8.	**AI CASE STUDIES**	**119**
	8.1 Predicting the performance of a warehouse	119
	8.2 The weather – concentration on data preparation	124
	8.3 A Cambridge spin-out start-up	130
	8.4 Summary	133
	REFERENCES	**134**
	FURTHER READING	**140**
	Machine learning	140
	Advanced theoretical text	140
	Index	141

FIGURES AND TABLE

Figure 1.1	Where AI sits compared to ML, NNs and deep learning	11
Figure 1.2	Dilts' logical levels of change	13
Figure 2.1	The learning agent – based on the structure of agents in Russell and Norvig	27
Figure 2.2	The state of the world – the agent's internal world that can make up for incomplete information or help to make decisions	29
Figure 2.3	Honda's ASIMO	33
Figure 2.4	Hierarchical paradigm in robotic design	34
Figure 2.5	Reactive paradigm in robotic design	35
Figure 2.6	Hybrid deliberative/reactive paradigm in robotic design	35
Figure 3.1	The three pillars of sustainability	38
Figure 4.1	A position vector, $\boldsymbol{p}$, made of up three scalars, x, y and z	52
Figure 4.2	Example problem for vector calculus	55
Figure 4.3	Complexity and randomness	57
Figure 4.4	Heuristics can simplify complexity and randomness	58
Figure 4.5	Venn diagrams for false (none of the space S is covered) and true (all of the space S is covered)	60
Figure 4.6	Venn diagrams for the probability of finding a letter A and the probability of finding a letter B; A and B are not mutually exclusive	61
Figure 4.7	Venn diagram where A and B are not mutually exclusive. Centre region is accounted for twice in the addition law of probability	61
Figure 4.8	Venn diagram defining the conditional probability of event A, given event B	63
Figure 4.9	Example of a probability mass function of rolling two dice	64
Figure 4.10	Example of a probability density function of continuous data	65
Figure 4.11	Uniform and Gaussian probability density functions of continuous data	66
Figure 4.12	An example of the central limit theorem using five dice and rolling set of scores	67
Figure 4.13	Supervised learning	71
Figure 4.14	Unsupervised learning	72
Figure 4.15	Semi-supervised learning	73
Figure 4.16	Reinforcement learning	74
Figure 4.17	A simple example of over-fitting	75
Figure 4.18	A simple example of under-fitting	76

FIGURES AND TABLE

Figure 4.19	Algorithm assessment, the trade-off and finding the balance between under- and over-fitting	77
Figure 4.20	Visualisation is learning from experience at every stage of an ML project	77
Figure 4.21	Standard data visualisation examples	79
Figure 4.22	Open-source software to build simulated environments – Blender	80
Figure 4.23	AI in medicine – humans and machines working together	81
Figure 4.24	The schematic of a neural network neuron	83
Figure 4.25	The schematic of a neural network with an input, hidden and output layer	84
Figure 8.1	Store's heat load: Vault One and Vault Two	123
Figure 8.2	Niagara Falls – the turbulence here dissipates enough energy to power an area the size of north Wales	126
Figure 8.3	The movement of the Earth's oceans	127
Figure 8.4	The visualisation of a large data set from a simulation of turbulence that is found in the Earth's atmosphere or a soap film	129
Figure 8.5	Optalysys's first optical mathematical operation was the derivative; the first prototype was built on a 250kg optical bed and was over a metre in length	132
Figure 8.6	Optalysys's first commercial optical AI processor – FT: X Optical co-processor system with PCIe drive electronics. It fits onto a standard PCI card about 100mm wide by 300mm long. The thumbnail, top left, is the next generation of AI processor and measures 32mm × 32mm	132
Table 5.1	Examples of the types of learning and algorithms	90

AUTHORS

Andrew Lowe is an engineer specialising in using computers to solve the challenges engineers face. He started as a nuclear design apprentice when computers were replacing drawing boards. Learning from experience has been a fundamental part of his career, and artificial intelligence helps him in his day-to-day work – some of the problems he has worked on can only be solved with supercomputers. He obtained his PhD from Cambridge and has worked in academia and industry. He has also helped an AI start-up company gain traction. He is married with two daughters, and lives and works in the Lake District, a World Heritage Site. He volunteers with local organisations to give people better opportunities.

Steve Lawless has always had a keen interest in science and technology, and has worked in the computer industry for over 40 years. He has trained literally thousands of people to give them the skills they need to make computers work. He knows Andy from volunteering at the local ski club. He runs a successful training company who have a worldwide customer base. He has written more than 100 training courses on IT, and enjoys making technology easy to understand and accessible. He is married with three sons and two daughters, and also lives and works in the Lake District.

ACKNOWLEDGEMENTS

We would like to thank our families, who have been patient. It is amazing how much time it takes to pull together learning from experience in an easy-to-understand book.

The technical side of AI can be challenging, so the numerous teachers, lecturers, supervisors, tutors and mentors are too many to mention individually. Thank you; you know who you are. Especially to those who have made their lectures and notes available on the web so everyone can learn.

We would like to thank Dr Paul Mort, of Sellafield Sites Ltd, who has been a keen enthusiast and supporter of the BCS AI Essentials and AI Foundation courses. His challenging questions while reviewing the courses made us think very carefully. Paul is a roboticist, and we were highly aware that AI is a much more enriching and broader subject than machine learning. Indeed, we have all left the digital revolution behind in the third industrial revolution, and we are now well on our way to the fifth. The AI agent is a really important and fundamental concept that we can easily overlook if we think AI is a digital machine learning technology.

BCS have played a big role in bringing this book and the courses together. We would like to thank, in no particular order, Ann Winskill, Felicity Page, Ian Borthwick, Rebecca Youé, Chris Leadbeater, Becky Lemaire, Sharon Pillai, Blair Melsom, Helen Silverman, Natalie Rew, Jo Massiah, Pam Fegan and Suky Sunner. There have been many others involved, we are sure. Thank you.

Many thanks go to Debbie Patten and Nigel Crowdy, at Purple Griffon, who make course organisation look easy. People who have attended the course have given highly valued contributions to the course as we have developed it. Learning from experience is exactly what we do on the courses; we value your feedback, creative input and challenging questions.

We are sure you will find something useful in this book, and hope that above all you are sceptical, critical and rigorous in your study and learn from experience. We are human, we make mistakes, and would love to hear all feedback, good and bad. This is part of the scientific method and learning from experience.

ABBREVIATIONS

ABM	agent-based modelling
AGI	artificial general intelligence
AI	artificial intelligence
AIoT	Artificial Intelligence of Things
ANN	artificial neural network
AR	augmented reality
ATEAC	Advanced Technology External Advisory Council Agency
CGI	computer-generated imagery
CNN	Convolutional Neural Network
CPU	central processing unit
DARPA	Defence Advanced Research Projects Agency
DBN	deep belief network
DNN	deep neural network
EU	European Union
GAN	generative adversarial network
GPU	graphics processing unit
IA	intelligent automation/augmentation
ICT	information and communications technology
IoT	Internet of Things
IT	information technology
k-NN	k-Nearest Neighbour
LSTM	long short-term memory
ML	machine learning
MLP NN	multilayer Perceptron neural network
NGO	non-governmental organisation
NI	natural intelligence
NLP	natural language processing
NN	neural network
NPC	non-player character
OCR	optical character recognition

ABBREVIATIONS

OPU	optical processing unit
PDF	probability density function
PESTLE	Political, Economic, Sociological, Technological, Legal and Environmental
R&D	research and development
RBM	restricted Boltzmann machine
RNN	recurrent neural network
RPA	robotic process automation
SDG	sustainable development goal
SFIA	Skills for the Information Age
SVM	support vector machine
TPU	tensor processing unit
TRL	Technology Readiness Level
UX	user experience
VR	virtual reality

GLOSSARY

Artificial intelligence (AI): Artificial intelligence refers to systems designed by humans that, given a goal, act in the physical world by perceiving their environment; to build intelligent entities.

Assistive robot: A robot designed to assist a person.

Autonomous system: A system that can make decisions autonomously.

Autonomy: The ability of an object to make its own decisions.

Axon terminals: Axon terminals are terminations of the telodendrions (branches) of an axon.

Backpropagation: A method of training the weights used in artificial neural networks.

Bayesian network: A Bayesian network, or belief network, is a probabilistic graphical model that represents a set of variables and their conditional dependencies.

Bias: Bias is the deviation of a statistical estimate from the actual quantity value. It can also mean the conscious and unconscious bias demonstrated in humans' behaviours.

Big Data: Big Data refers to data sets that are so big and complex that traditional data-processing application software are inadequate to deal with them.

Boosting: Boosting is an ensemble algorithm that aims to reduce bias, and to convert weak learners to strong ones.

Bootstrap aggregating – bagging: Bootstrap aggregating is an ensemble algorithm used in classification and regression.

Chatbot: A chatbot is an artificial intelligence program that conducts a conversation via auditory or textual methods.

Classification: Classification is the problem of identifying to which set of classes an object belongs.

Clustering: Clustering groups a set of objects in such a way that objects in the same group are similar to each other.

Cognitive simulation: Cognitive simulation uses computers that test how the human mind works.

Combinatorial complexity: Combinatorial complexity is the exponential growth in computer power required to solve a problem that has many ever-increasing complex combinations.

Combinatorial explosion: A combinatorial explosion is the rapid growth of the number of combinations a system has to deal with.

Connectionist: Cognitive science that hopes to explain intellectual abilities using artificial neural networks.

Data analytics: Meaningful patterns are found from data via discovery and interpretation.

Data cleaning: Data cleaning prepares data for analysis.

Data mining: The process of discovering patterns in large data sets.

Data science: Data science uses scientific methods, processes, algorithms and systems to understand data.

Data scrubbing: See data cleaning.

Decision trees: A decision tree uses a tree-like graph or model of decisions.

Deep learning: Deep learning is a multi-layered neural network.

Dendrites: Dendrites are branched extensions of a nerve cell that propagate the electrochemical stimulation.

Edges: Edges are the geometric machine learning name for the brain's axons.

Emotional intelligence or emotional quotient (EQ): The understanding of our emotions and the emotions of others.

Ensemble: Ensemble methods use multiple or a combination of learning algorithms to obtain better learning outcomes.

Ethical purpose: Ethical purpose is used to indicate the development, deployment and use of AI that ensures compliance with fundamental rights and applicable regulation, as well as respecting core principles and values. This is one of the two core elements to achieve trustworthy AI.

Expert system: A computer system that emulates the decision-making ability of a human expert.

Feedforward neural network: A feedforward neural network is an artificial neural network that iterates to find the weights of the network from information passing from the input to the output.

Fourth industrial revolution: The fourth industrial revolution represents new integrated approaches where technology becomes embedded within objects and societies.

Functionality: The tasks that a computer software program is able to do.

Genetic algorithms: A genetic algorithm is an algorithm inspired by the process of natural selection.

Hardware: Hardware are the physical parts or components of a computer.

Heuristic: Heuristic is a strategy derived from previous experiences with similar problems.

High-performance computing – supercomputing: A computer with very high performance.

Human-centric AI: The human-centric approach to AI strives to ensure that human values are always the primary consideration. It forces us to keep in mind that the development and use of AI should not be seen as a means in itself. The goal of increasing citizens' wellbeing is paramount.

Hyper-parameters: Hyper-parameters, set before the learning process begins, are parameters that tune the algorithms' performance in learning.

Inductive reasoning: Inductive reasoning makes broad generalisations from specific observations.

Intelligent quotient (IQ): A standard test of intelligence.

Internet of Things (IoT): IoT is the network of devices. Devices have embedded technology and network connectivity.

k-means: k-means is a clustering algorithm that generates k clusters. An object belongs to a cluster with a nearest mean to a prototype.

k-Nearest Neighbour (k-NN): The simplest clustering algorithm.

Layers: Neural networks are made up of multiple layers. Each layer is made up of an array of nodes.

Linear algebra: The branch of mathematics concerning linear equations and functions and their representations through matrices and vector spaces.

Logistic regression: Used in classification to predict the probability of an object belonging to a class.

Machine learning (ML): Machine learning is a subset of artificial intelligence. This field of computer science explores a computer's ability to learn from data.

Model optimisation: The improvement of the output of a machine learning algorithm (e.g. adjusting hyper-parameters).

Natural language processing (NLP): An area of artificial intelligence concerned with the interactions between computer and human (natural) languages.

Natural language understanding: A term used to describe machine reading comprehension.

Nearest neighbour algorithm: The nearest neighbour algorithm was one of the first algorithms used to find nearest neighbours between sets of data (e.g. the travelling salesman problem).

Neural network (NN): A machine learning algorithm based on a mathematical model of the biological brain.

Nodes: Nodes represent neurons (biological brain) and are interconnected to form a neural network.

One-hot encoding: Transforms object features into a numerical form for use in algorithms (e.g. false is given the number 0 and true is given the number 1).

Ontology: The philosophical study of the nature of objects' being and the relationships of objects.

Optical character recognition (OCR): The conversion of images of typed, handwritten or printed text into a form a computer can use.

Over-fitting or over-training: Over-fitting is a machine learning model that is too complex, has high variance and low bias. It is the opposite of under-fitting or under-training.

Probabilistic inference: Probabilistic inference uses simple statistical data to build nets for simulation and models.

Probability: Probability is the measure of the likelihood that an event will occur.

Pruning: Pruning reduces the size of decision trees.

Python: A programming language popular in machine learning.

Random decision forests or random forests: Random decision forests are an ensemble learning method for classification, regression and other tasks.

Regression analysis: In machine learning, regression analysis is a simple supervised learning technique used to find a trend to describe data.

Reinforcement (machine) learning: Reinforcement learning uses software agents that take actions in an environment in order to maximise some notion of cumulative reward or minimise a loss.

Robotic process automation: More often known as RPA, robotic process automation is a business process automation technology based on the notion of software robots.

Robotics: Robotics deals with the design, construction, operation and use of robots.

Scripts: Scripts are programs written in an application's run-time environment. They automate tasks, removing the need for human intervention.

Search: The use of machine learning in search problems (e.g. shortest path, adversarial games).

Semi-supervised machine learning: Machine learning that uses labelled and unlabelled data for training.

Sigmoid function: A Sigmoid function is a mathematical function. It is a S-shaped, or Sigmoid, curve and is continuous.

Software: A generic term for the instructions that make a machine deliver functionality.

Software robots: A software robot replaces a function that a human would otherwise do.

Strong AI or artificial general intelligence (AGI): Strong AI's goal is the development of artificial intelligence with the full intellectual capability of a human and not just to simulate thinking.

Supervised machine learning: Supervised machine learning uses labelled data to map an input to an output.

Support vector machine: A support vector machine constructs a hyperplane or set of hyperplanes (a linear subspace or subspaces).

Swarm intelligence: Swarm intelligence is the collective behaviour of self-organised systems, natural or artificial.

Symbolic: Symbolic artificial intelligence is the term in artificial intelligence research based on high-level 'symbolic' (human-readable) representations of problems.

System: A regularly interacting or interdependent group of objects that form a whole.

Trustworthy AI: Trustworthy AI has two components: (1) its development, deployment and use should comply with fundamental rights and applicable regulation as well as respecting core principles and values and ensuring ethical purpose; (2) it should be technically robust and reliable.

Turing machine: A mathematical model of computation.

Under-fitting or under-training: Under-fitting is when the machine learning model has low variance and high bias. It is the opposite of over-fitting or over-training.

Universal design: Universal design (a close relation to inclusive design) refers to broad-spectrum ideas meant to produce buildings, products and environments that are inherently accessible to all people.

Unsupervised machine learning: Machine learning that learns a function from unlabelled data.

Validation data: A set of data used to test the output of a machine learning model that is not used to train the model.

Variance: The expectation of the squared deviation of a random variable from its mean.

Visualisation: Any technique for creating images, diagrams or animations to communicate a message.

Weak AI or narrow AI: Weak artificial intelligence (weak AI), also known as narrow AI, is artificial intelligence focused on one narrow task. It is the contrast of strong AI.

Weights: A weight is a mathematical object used to give an object or objects additional/diminished influence or weight.

USEFUL WEBSITES

ARTIFICIAL INTELLIGENCE

European Defence Agency, European Defence Matters journal
https://www.eda.europa.eu/webzine/issue14/cover-story/artificial-intelligence-(ai)-enabled-cyber-defence

European Parliament Committees, Hearings on artificial intelligence
https://www.europarl.europa.eu/committees/en/indexsearch?scope=CURRENT&term=9&query=artificial+intelligence&scope=ALL&ordering=RELEVANCE

Open AI
https://openai.com/

ROBOTS

Skills for Care, Scoping study on the emerging use of AI and robotics in social care
https://www.skillsforcare.org.uk/Documents/Topics/Digital-working/Robotics-and-AI-in-social-care-Final-report.pdf

ETHICS

Alan Turing Institute, Data ethics: how can data science and artificial intelligence be used for the good of society?
https://www.turing.ac.uk/research/data-ethics

European Union, Ethics guidelines for trustworthy AI
https://ec.europa.eu/digital-single-market/en/news/ethics-guidelines-trustworthy-ai

Future of Life Institute
https://futureoflife.org/

MACHINE LEARNING

Google for Education, Google's Python class
https://developers.google.com/edu/python

The Royal Society, What is machine learning?
https://royalsociety.org/topics-policy/projects/machine-learning/

SUSTAINABILITY

International Organization for Standardization
https://www.iso.org/home.html

SmartCitiesWorld
https://www.smartcitiesworld.net/

United Nations, Take action for the sustainable development goals
https://www.un.org/sustainabledevelopment/sustainable-development-goals/

PREFACE

This book was written with the express purpose of supporting the BCS AI Essentials and the BCS AI Foundation training courses and other scheduled courses already under development in the BCS AI course pipeline at the time of writing.

Its aim is to document what artificial intelligence is and what it is not, separate fact from fiction and educate those with an interest in AI. We have also included a number of topics that introduce the basics of machine learning and ethics.

We believe that this book is unique in that it brings together information and concepts that until now have been spread across numerous other volumes. The book also aims to simplify (where possible) complex and confusing AI concepts, making the topics highly accessible to those without a high-level degree in the subjects covered.

Our aim here is to bring these concepts to life by balancing theory with practice. We want to make the human part of an AI project as important as the AI itself. After all, machines are here to take the heavy lifting away from us humans. Not only that, but to give us extra capabilities that we wouldn't have by ourselves. Humans and machines have unique capabilities, and it is important to find the right balance between them.

People, society and governments are quite rightly concerned with AI and its potential. As such, we have adopted the EU guidelines (https://ec.europa.eu/digital-single-market/en/news/ethics-guidelines-trustworthy-ai) on the ethical use of AI. These guidelines ask us to build human-centric ethical purpose that gives us trustworthy and technically robust AI. This puts us humans at the goal setting lead in AI.

Stuart Russell's[1] recent book on human compatible AI puts the human into our consideration when undertaking an AI project. His take on this asks if we should think about AI as serving our needs, telling us how to be better humans. In doing so it gives us an alternative to controlling AI, by asking the AI what is best for us as humans.

As we move into the fifth industrial revolution, we have the opportunity to think about humans and machines. How do we complement each other? How can AI and machines leave humans to undertake more valued work and deal with ambiguous or contradictory situations, to become more human, to build better societies and for all to exploit their talents? What are the new roles for humans, humans and machines, and machines only? Simply considering these roles and focusing on opportunities paves the way for a richer environment as we progress. We focus on our needs and are less distracted by the notion of robots coming to take over our jobs! Can you, your current field or organisation benefit from learning from experience? If so, read on.

1 INTRODUCTION: ETHICAL AND SUSTAINABLE HUMAN AND ARTIFICIAL INTELLIGENCE

This chapter sets the scene for artificial intelligence (AI). We look at intuitive definitions of human and artificial intelligence. We also introduce the European Union's (EU's) ethical guidelines for AI and take a look back at the progress of AI over the past couple of centuries, examining how we as humans relate to this disruptive technology.

1.1 THE GENERAL DEFINITION OF HUMAN INTELLIGENCE

It is a vast understatement to say that human beings are one of the wonders of the universe. Human intelligence is the culmination of billions of years of evolution from single cell organisms to what we are today, which is ultimately marked by our ability to undertake complex mental feats and be self-aware. It also includes the ability to recognise our place in the universe and ask annoying philosophical questions such as 'Why are we here?' and 'What is our purpose?'

There are many definitions of human intelligence. Our chosen definition is useful because it is intuitive and gives us a practical base that builds a strong foundation for AI. In fact, it needs to be a little more than intuitive: it also needs to guide us as to what AI is useful for in practice. When considering this definition, we must keep in the back of our minds that the need to find the right balance of theory and practice is paramount. We will also need to understand that AI and machine learning (ML) have significant limitations, and this will become apparent as we moved through the book.

Take five minutes to think about what it means to be human.

Leonardo da Vinci captured the essence of his science in art.[2] René Descartes stated '*cogito, ergo sum*' (I think therefore I am).[3] Ada Lovelace wrote the first algorithm and notes on the role of humans and society with technology.[4] Neil Armstrong was the first to put one foot on the moon, and changed our perspective of the world completely.[5] Roger Bannister was the first to run a mile in under four minutes. Dr Karen Spärck Jones gave us the theoretical foundations of the search engine.[6] Tu Youyou is a tenacious scientist who discovered a cure for malaria, which won her the Nobel Peace Prize in 2015.[7] Tu's intellectual talents are amazing, and, after perfecting her cure, she volunteered to be the first person for it to be tested on. We could ask ourselves if this was confidence or bravery.

We have set ourselves up here to introduce the concept of subjectivity. We have free will and all of us have our own unique subjective experience. We are conscious and subjective, and conscious experience is something that we will need to be all too aware of as we develop AI.

> We must always consider what effects artificial intelligence will have on humans and society.

Robert Sternburg gives us a useful definition of human intelligence,[8] at least in so far as it relates to AI:

> Human intelligence: mental quality that consists of the abilities to learn from experience, adapt to new situations, understand and handle abstract concepts, and use knowledge to manipulate one's environment.

Here we can quickly recognise the desire to manipulate our environment in some way, and that we will use all our human talents to do so. It is general, and we still need to identify the type of learning from experience that humans do, or, to put it another way, what machines can help us with.

Sometimes called natural intelligence (NI), human intelligence is generally considered to be the intellectual accomplishment of humans and has been discussed by philosophers for thousands of years. Of course, other living things possess NI to some degree as well, but for now let's just consider human intelligence. We are biased, but it's fair to say that humans are the most intelligent living organisms on the planet. Just as we developed tools in the Stone and Iron Ages, we are now equipping ourselves with machines to help us intellectually.

We may phrase this as coming to the correct conclusions – hypothesising and testing – and understanding what is real, although we sometimes get it wrong. It's also about how to understand complex problems like weather prediction or winning a game of chess; adapting what we have learned through things like abstraction, induction, simplification and creativity. It allows us to adapt and control our environment and interact socially, giving us an evolutionary advantage.

It may also make sense to consider human intelligence from a number of perspectives, such as:

- **Linguistic intelligence** – the ability to communicate complex ideas to another.
- **Mathematical intelligence** – the ability to solve complex problems.
- **Interpersonal intelligence** – the ability to see things from the perspective of others, or to understand people in the sense of having empathy.

So, how do we acquire these particular skills or traits? Through learning from **experience**.

1.1.1 Human learning

Human learning is the process of acquiring knowledge. It starts at a very early age, perhaps even before we are born. Our behaviour, skills, values and ethics are acquired and developed when we process information through our minds and learn from those experiences. Human learning may occur as part of education, personal development or any other informal/formal training, and is an ongoing process throughout our lives. Each person has a preference for different learning styles and techniques (e.g. visual, aural, kinaesthetic, etc.).

> Machine learning can give us super-human capability; we can search every research paper using a search engine on our smartphone. This could take 'old school' academics a lifetime of hard work, travel and focused attention. Machine learning is changing our beliefs, behaviour and speed of progress.

Now we understand the basics of **human intelligence** and **human learning**, let's see how that compares to **artificial intelligence** and **machine learning**. We will also explain further some of the jargon that is thrown around in AI circles and explain exactly what artificial intelligence is – and what artificial intelligence isn't.

1.2 DEFINITION OF ARTIFICIAL INTELLIGENCE

In simple terms, AI is intelligence demonstrated by machines, in contrast to the NI displayed by humans and other animals.

Stuart Russell and Peter Norvig, the authors of the standard AI textbook *Artificial Intelligence: A Modern Approach*,[9] explain that AI is a universal subject and helpful to us all. Learning from experience is AI's signature. We will use this concept a lot; it applies to machines as it does, perhaps more so, to humans.

Einstein is often quoted as saying: 'The only source of knowledge is the experience'.

Ask yourself the question: Can machines help us to learn from experience? If the answer to this question is yes, then AI can help you!

1.2.1 Artificial general intelligence (AGI)

AGI is the hypothetical intelligence of a machine that has the capacity to understand or learn any intellectual task that a human being can understand or learn. There is wide agreement among AI researchers that to do this AGI would need to perform a full range of human abilities, such as using strategy, reasoning, solving puzzles, making judgements under uncertainty, representing knowledge (including common-sense knowledge), planning, learning and communicating in natural language, and integrate all these skills towards common goals. AGI might never be achieved, and not everyone agrees whether it is possible or if we will ever get there.

A number of tests have been put forward to decide if and when AGI (human-like intelligence) has been achieved. One of the first tests was the 'Turing' test devised by the British scientist Alan Turing. The 'Turing' test goes along the lines of a machine and a human conversing while heard but unseen by a second human (the evaluator), who must evaluate which of the two is the machine and which is the human. The test is passed if they can fool the evaluator a significant fraction of the time. Turing did not, however, prescribe what should qualify as intelligence. Several other tests have since been defined, including visual and construction tests. In reality, a non-human agent would be expected to pass several of these tests.

Current AGI research is extremely diverse and often pioneering in nature, and estimates vary from 10 to 100 years before AGI is achieved. The consensus in the AGI research community seems to be that the timeline discussed by Ray Kurzweil in *The Singularity is Near*[10] (i.e. between 2015 and 2045) is plausible. Kurzweil has based his estimate of 2045 on the exponential advances in four key areas of research: AI, robotics, genetic engineering and nanotechnologies.

There are other aspects of the physical human brain besides intelligence that are relevant to the concept of strong AGI. Strong AGI is AGI with consciousness. We discuss consciousness in more depth later in the book. If strong AI or some kind of conscious AI emerges in the future, then our ethical development of these technologies is paramount. Ethical use of AI is a fundamental requirement, and we need to build in our ethics from the start. It's not too late.

1.3 A BRIEF HISTORY OF ARTIFICIAL INTELLIGENCE

Today we are mainly concerned with the current use and future applications of AI, and the benefits we would like to obtain from its use. But we should not ignore where and when AI first appeared, the history of AI and the challenges encountered along the way. To ignore our AI history would be to risk some of those challenges recurring.

Before we look at AI's history, it's worth noting that it is now generally recognised that John McCarthy coined the term 'artificial intelligence' in 1955. John McCarthy is one of the 'founding fathers' of AI, together with Alan Turing, Marvin Minsky, Allen Newell and Herbert A. Simon.

1.3.1 Back in antiquity

Way back in antiquity there were legends, stories and myths of effigies and artificial mechanical bodies endowed with some form of intelligence, typically the creation of a wise man or master craftsman. We still have references to golems of biblical times, which were magical creatures made of mud or clay and brought to 'life' through some incantation or magic spell.

Aristotle (384–322 BC), the Greek polymath and father of later Western philosophy, was the first to write about objects and logic and laid the foundations of ontology and the scientific method. As a result, today we teach natural science, data science, computer science and social science.

Without the required technology, many centuries passed without any progress in the pursuit of AI.

1.3.2 The 18th and 19th centuries

In the 18th century we saw the mathematical development of statistics (Bayes theorem) and the first computer description and algorithm from Ada Lovelace.

During the 19th century, AI entered the world of science fiction literature with Mary Shelley's *Frankenstein* in 1818 and Samuel Butler's novel *Erewhon* in 1872, which drew on an earlier (1863) letter he had written to *The Press* newspaper in New Zealand, 'Darwin among the Machines'. Butler was the first to write about the possibility that machines might develop consciousness by natural selection.

In 1920 Czech-born Karel Čapek introduced the word 'robot' to the world within his stage play, *R.U.R.* (Rossum's Universal Robots). 'Robot' comes from the Slavic language word *robota*, meaning forced labourer.

AI has since become a recurrent theme in science fiction writing and films, whether utopian, emphasising the potential benefits, or dystopian, focusing on negative effects such as replacing humans as the dominant race with self-replicating intelligent machines. To some extent, we can say that what was yesterday's science fiction is quickly becoming today's science fact.

1.3.3 Technological advances during the 'golden years' (the second half of the 20th century)

In 1943 Warren McCulloch and Walter Pitts created a computational model for neural networks (NNs), which opened up the subject. The first was an electronic analogue NN built by Marvin Minksy. This sparked the long-lasting relationship between AI and engineering control theory.

Then in 1950 the English mathematician Alan Turing published a paper entitled 'Computing Machinery and Intelligence' in the journal *Mind*. This really opened the doors to the field that would be called Artificial Intelligence. It took a further six years, however, before the scientific community adopted the term 'artificial intelligence'.

John McCarthy organised the first academic conference on the subject of AI at Dartmouth, New Hampshire, in the summer of 1956. The 'summer school' lasted eight weeks and brainstormed the area of artificial intelligence. It was originally planned to be attended by 10 people; in fact, people 'dropped in' for various sessions and the final list numbered nearly 50 participants. Russell and Norvig quoted McCarthy's proposal for the summer school (p. 17):[9]

> We propose that a 2-month, 10-man study of artificial intelligence be carried out during the summer of 1956 at Dartmouth College in Hanover, New Hampshire. The study is to proceed on the basis of the conjecture that every aspect of learning or any other feature of intelligence can in principle be so precisely described that a machine can be made to simulate it. An attempt will be made to find how to make

machines use language, form abstractions and concepts, solve kinds of problems now reserved for humans, and improve themselves. We think that a significant advance can be made in one or more of these problems if a carefully selected group of scientists work on it together for a summer.

The mid-1950s into the early 1960s also saw the start of machines playing draughts, checkers and chess. This was the start of 'Game AI', which is still big business to this day and has developed into a multi-billion-dollar industry.

In 1959, Arthur Lee Samuel first used and popularised the term 'machine learning', although, today, Tom Mitchell's definition is more widely quoted (see Section 1.5). Samuel's checkers-playing program was among the world's first successful self-learning programs, and was amazing given the limited technology available to him at the time.

The 1960s saw the development of logic-based computing and the development of programming languages such as Prolog. Previously it took an astronomical number of calculations to prove simple theorems using this logic method. This opened up a debate regarding how people and computers think.

Hubert Dreyfus, a professor of philosophy at the University of California Berkeley (1968–2016), challenged the field of AI in a book,[11] explaining that human beings rarely used logic when they solved problems. McCarthy was critical of this argument and the association of what people do as being irrelevant to the field of AI.[12] He argued that what was really needed were machines that could solve problems – not machines that think as people think.

In 1973 many funding resources were withdrawn from AI research, mainly as a result of Sir James Lighthill's report into the state of AI, and ongoing pressure from US Congress and the US and British governments.[13] As a result, funding was reduced for research into artificial intelligence, and the difficult years that followed would later be known as an 'AI winter', lasting from 1974 to 1980. Sir James highlighted that other sciences could solve typical AI problems; AI would hit combinatorial explosion limits and practical problems would resign AI to solving only trivial toy problems. In his view, general AI could not be achieved and there was no prospect of a general AI robot ever.

During the 1970s, however, a number of breakthroughs were made, most notably in 1974 by Ray Kurzweil's company, Kurzweil Computer Products Inc, developing optical character recognition (OCR) and a text-to-speech synthesizer, thus enabling blind people to have a computer read text to them out loud. It was unveiled to the public in 1976 at the National Federation for the Blind, and became commercially available in 1978. Kurzweil subsequently sold the business to Xerox. It is widely considered to be the first AI product, although today we don't associate OCR with AI or ML because it is now routine and we take it for granted. It was the precursor to reading handwritten text and the development of natural language processing (NLP).

Following on from the 'AI winter', the 1980s saw a boom in AI activity. In 1986 David Rumelhart and James McClelland developed ideas around parallel distributed processing and neural network models. Their book, *Parallel Distributed Processing: Explorations in the Microstructure of Cognition*, described their creation of computer simulations of perception, giving computer scientists their first testable models of neural processing.[14]

The 1980s also saw the rise of the robots, with many researchers suggesting that AI must have a body if it is to be of use; it needs to perceive, move, survive and deal with the world. This led to developments in sensor-motor skills development. AI also began to be used for logistics, data mining, medical diagnosis and other areas.

During the 1990s a new paradigm called 'intelligent agents' became widely accepted in the AI community. An intelligent agent is a system that perceives its environment and takes actions that maximise its chances of success.

In 1997, IBM's Deep Blue became the first computer chess-playing system to beat a reigning world chess champion, Garry Kasparov. It won by searching 200,000,000 moves per second. By comparison, 20 years on, Apple's iPhone7 was 40 times faster than Deep Blue had been in 1997.

1.3.4 Bringing us up to date: the first 20 years of the 21st century

Over the last 20 years, developments in technology with cheaper and faster computers have finally caught up with our AI aspirations. We have started to gain the computing power to really put AI to work.

In 2002 iRobot released Roomba, which autonomously vacuums a floor while navigating and avoiding obstacles. It sold a million units by 2004, and over 8 million units by 2020. iRobot then went on to create a range of other commercial, environmental, military and medical robots.

In 2004 the Defence Advanced Research Projects Agency (DARPA), a prominent research organisation of the United States Department of Defense, introduced the DARPA Grand Challenge, offering prize money for competitors to produce vehicles capable of travelling autonomously over 150 miles. Then in 2007, DARPA launched the Urban Challenge for autonomous cars to obey traffic rules and operate in an urban environment, covering 60 miles within six hours.

Google entered the self-driving autonomous market in 2009 and built its first autonomous car, which sparked a commercial battle between Tesla, General Motors, Volkswagen and Ford, to name a few entrants into the same market.

From 2011 to 2014 a series of smartphone apps were released that use natural language to answer questions, make recommendations and perform actions: Apple's Siri (2011), Google's Google Now (2012) and Microsoft's Cortana (2014).

SCHAFT Inc of Japan, a subsidiary of Google, built robot HRP-2, which defeated 15 teams to win DARPA's Robotics Challenge Trials in 2013. The HRP-2 robot scored 27 out of 32 points over eight tasks needed in disaster response. The tasks were to drive a vehicle, walk over debris, climb a ladder, remove debris, walk through doors, cut through a wall, close valves and connect a hose.

In 2015, an open letter petitioning for the ban in development and use of autonomous weapons was signed by leading figures such as Stephen Hawking, Elon Musk, Steve Wozniak and over 3,000 researchers in AI and robotics.

In 2016 Google's DeepMind AlphaGo supercomputer beat Lee Se-dol, a world Go champion, at a five-game match of Go (a strategic board game that originated in China more than 2,500 years ago and is considered one of the most complex strategy games in the world). It took just 30 hours of unsupervised learning for the supercomputer to teach itself to play Go. Lee Se-dol was a 9 dan professional Korean Go champion who won 27 major tournaments from 2002 to 2016. He announced his retirement from the game in 2019, declaring that AI has created an opponent that 'cannot be defeated'.[15] In 2017 AlphaGo Zero, an improved version of AlphaGo, beat the world's best chess-playing computer program, StockFish 8, winning 28 of the 100 games and drawing 72 of them. What is astonishing is that AlphaGo Zero taught itself how to play chess in under four hours.

Also in 2017, the Asilomar Conference on Beneficial AI was held near Monterey, California. Thought leaders in economics, law, ethics and philosophy spent five days in discussions dedicated to beneficial AI. It discussed AI ethics and how to bring about beneficial AI while at the same time avoiding the existential risk from AGI.

1.3.5 The industrial revolutions

We are currently in the middle of the fourth industrial revolution and some would argue that we have already reached the fifth, although that has yet to be defined. In each industrial revolution, mankind has designed and developed technologies that have made a paradigm shift in human capabilities and exploited those technologies to drive through progress, although some would argue that not all of mankind benefited as a result of these advancements.

The first industrial revolution, sometimes referred to as the Industrial Revolution, occurred during the 18th and 19th centuries, primarily in Europe and the United States. It mainly grew between 1760 and 1840 (although the exact dates are open to debate) and resulted in the transition from hand production methods to machine production – initially led by the development of the steam engine powering large textile factories. It was a major turning point in history, led to worldwide trading and rapid population growth and resulted in large swaths of rural societies becoming urbanised and industrial.

The second industrial revolution occurred between 1870 and 1914, primarily across Europe, the United States and Japan. Sometimes known as the Technological Revolution, again it was a paradigm shift for mankind with the introduction of mass production and assembly lines. Increased use of electricity allowed for advancements in manufacturing and production, and resulted in technological advances such as the internal combustion engine, the telephone and the light bulb.

The third industrial revolution started in the 1950s and is often known as the Digital Revolution. It brought us space exploration, biotechnology, semiconductors, mainframe computing and information and communications technology (ICT), and embedded technology into society with personal computers, the internet and automated production of goods.

There is a bit of a blur between the third and fourth industrial revolutions. We believe that we are in the fourth today. The fourth exploits the gains made in the 'digital revolution' and is disruptive, driven by AI, robotics, Internet of Things (IoT), three-dimensional (3D) plastic printing, nanotechnology, bioengineering and so on.

I know what you're thinking: 'What happened between the industrial revolutions?' We didn't stop inventing or improving between the revolutions, and different parts of the globe experienced them at different times at different speeds, but these particular periods were paradigm shifts in thinking and invention.

The Diffusion of Innovation theory, developed by E. M. Rogers in 1962, explains how, over time, an idea or product gains momentum and spreads (diffuses) through a specific population, and he classifies adopters of innovations into five adopter categories: innovators, early adopters, early majority, late majority and laggards. This is based on the idea that certain individuals are inevitably more open to adoption and adaption than others. The adoption of AI technologies globally will be faster than the adoption of other technologies because we have seen a reduction in the time an industrial revolution lasts, from centuries to decades and now fractions of a decade, but we will still be led by innovators and there will still be laggards and even Luddites.

Human intelligence has led us through the various industrial revolutions. We are in the middle of the fourth, but what does this mean if we think about AI? It means increasingly that we will have more robots doing routine monotonous, laborious and dangerous tasks – doing the 'heavy lifting'. This introduces the idea of humans and machines working together at what they are each good at.

1.3.6 AI as part of 'universal design'

The concept of universal design was coined by the architect Ronald Mace, and is the design and composition of an environment so that it can be accessed, understood and used to the greatest extent possible by all people. For example, door handles, elevator controls and light switches should be designed for use by all people regardless of their age, size, ability or disability. A consideration, therefore, for any new AI service, system or product is that it is designed for all.

Incorporating the potential of AI in universal design can allow someone who is blind to ask a 'home assistant' what the weather is like, or for someone who is physically incapacitated to turn on the heating, or someone who is travelling home to turn on the heating while travelling.

A human working alongside an intelligent AI-enabled machine has the capability to do a lot more, whether that is in a work environment reducing physical risk or exertion, or on a personal basis educating us or translating our conversations. AI systems have the potential to make us more human.

Our efforts and endeavours developing AI-enabled products, systems and services should be focused on allowing us to be more human, improving us as humans (improving our physical and mental performance or by making us more active) or improving our ability to communicate or socialise.

The continual emergence of AI systems and products means that we as individuals and as part of wider society are going to have to reimagine every area of our lives to use AI in a positive way for all.

1.3.7 The concept of intelligent agents

A computer scientist may view AI as 'intelligent agents' perceiving their environment and taking actions to achieve a goal. Russell and Norvig describe intelligent agents in more detail.[9]

As we mentioned previously, the scientific method has allowed humans to develop at an ever quicker pace, and we can, broadly speaking, think of these as the industrial revolutions. From the 1980s AI has also adopted the scientific method[9] and in doing so has been absorbed into the revolutions. The basic intelligent agent can be very rudimentary and it's often difficult to see why they would be considered intelligent. Learning from experience is a common phrase used in AI and the learning agent is an intelligent agent that learns from experience.

The learning agent, as proposed by Russell and Norvig,[9] is an agent with an explicit ability to learn, to be autonomous. It is useful to always relate what we are doing in AI to the learning agent that can perceive its environment and take actions to achieve a goal. It also gives us an intuitive insight into what artificial intelligence and machine learning are – for instance to understand that ML is about learning from data – but that this alone is insufficient when we are designing products and services that will certainly operate in an environment.

> We can define other types of agent, but these may not learn. We explore this further in Section 2.1, Understanding the AI intelligent agent.

Deep learning is a technique that gives learning agents the ability to learn from sensors and actuators and has been very successful in achieving very complex tasks.

Figure 1.1 shows how deep learning fits into the overall schematic of AI.

1.3.8 Consciousness – the unsolved problem

Human consciousness, sometimes referred to as sentience, is, in its simplest terms, having an awareness of an internal or external existence or having a mental state you are aware of being in. This can be compared to subconsciousness, which is that part of your mind that notices and remembers information when you are not actively trying to do so and can often influence your behaviour even though you do not realise it.

Some humans fear conscious machines, however unlikely they may be. Consciousness is a complex area at the cutting edge of AI research; our knowledge is growing but we might never understand what consciousness actually is. Interfacing the human

Figure 1.1 Where AI sits compared to ML, NNs and deep learning

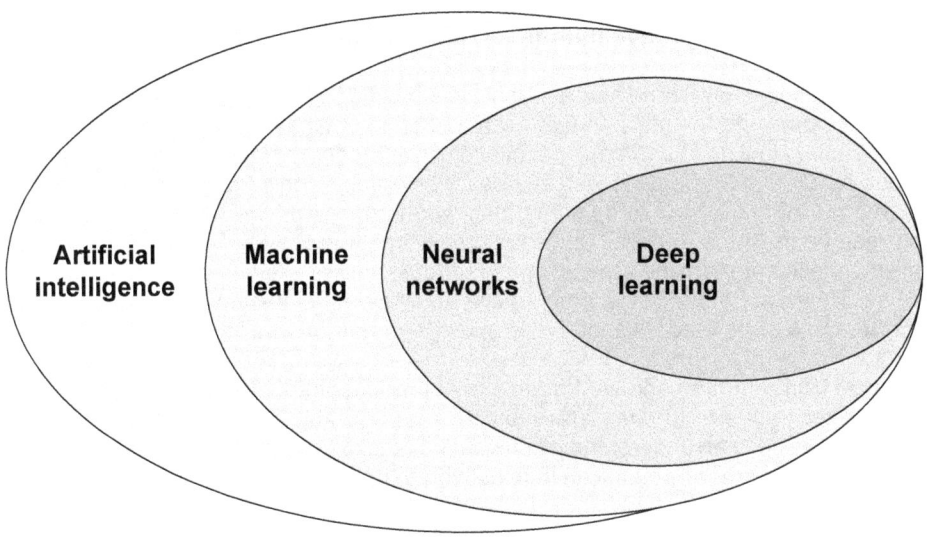

mind with machines could advance medical success or outcomes – examples that are currently being explored include stroke recovery.

> The most influential modern physical theories of human consciousness have been proposed by numerous neuroscientists, physicists and philosophers over the last couple of centuries, and they seek to explain consciousness in terms of neural events occurring within the brain.
>
> Nobody currently knows where consciousness actually comes from, but it's thought to be an emergent property that comes from neurons connected in a specific but complex way. We do understand how parts of our brain function, but much is still unknown. Interestingly, the study of organoids (e.g. small blobs of brain tissue grown in the laboratory) is allowing scientists to gain an understanding of how parts of the brain function.

Machine consciousness or artificial consciousness is the domain of cognitive robotics and is primarily concerned with endowing an artificial entity with intelligent behaviour by providing it with a processing architecture. This will allow it to learn and reason about how to behave in response to complex goals in a complex world. The aim of the theory of artificial consciousness is to define what would have to be synthesised to develop artificial consciousness in an engineered artefact such as a robot.

Many organisations are pursuing developments in artificial consciousness, but comparing consciousness in functional machines to consciousness in functional humans is more difficult than expected and the topic has raised a debate around the risks of developing conscious machines.

1.3.9 Modelling subjective humans – neuro-linguistic programming

By understanding how some humans achieve amazing things, can we then teach others to achieve amazing things? Can we teach a machine? What makes a good astronaut, for instance? How do charismatic people convince others to follow them? What makes a good marketing strategy? If we can't describe – and by this we mean explicitly – or teach other humans how to do something, what chance do we have to design, build and teach a machine to do it? Our source of intellect is the human. The role models for any intelligent machine are currently humans, but we are subjective and objective beings. The human intellect is much more than scientific objectivity; neuro-linguistic programming gives readers another perspective of human intelligence.

AI learns from humans and we humans give the machines goals, environments and learning techniques. Humans are subjective beings and, if we are going to teach machines, we need an understanding of what it takes to understand subjective humans. Neuro-linguistic programming can help us with that.

Practically, we will also need to champion our AI, especially if AI is new to an organisation. This means dealing with the subjective nature of management, stakeholders and people in your team. We will need to capture hearts and minds in order to champion our AI projects. Neuro-linguistic programming is also a useful approach in this regard; it has been adopted by some of the world's most successful organisations. This section is quite general, but vital because, in developing or championing your AI projects, you will need these skills.

Neuro-linguistic programming asks: 'Are the processes that humans use to achieve amazing things the same as the processes that humans use to hold them back?' Sometimes when humans are held back, this can be negative and destructive.

> Neuro-linguistic programming comes from the work of John Grinder and Richard Bandler, presented in their book, *The Structure of Magic*.[16] They diligently studied how skilled psychologists counselled humans, and realised that we humans do the following:
>
> - generalise;
> - delete;
> - distort.

Neuro-linguistic programming is now a creative, exciting array of tools, techniques and approaches to understand how humans and organisations achieve amazing things. However, it is not possible to cover neuro-linguistic programming in depth in a book on AI; instead, we'll give an overview of a couple of fundamental points taught to practitioners. The objective here is to understand our AI role models – humans – well and be able to handle the challenges of running an AI project and the subjective parts of the project specifically, such as motivating a team, dealing with the fears of individuals and organisations, creating a culture supporting AI, building business cases, communicating and influencing stakeholders. Neuro-linguistic programming could also play a role in how humans and machines interact in the future.

Our first neuro-linguistic programming example is building rapport, something that we need to be able to do well. One of the ways we might do this is understanding the type of person we are communicating with. Are they, for example, visual and express themselves in terms of visual words, such as 'I see what you want to achieve', 'I saw red the moment I read that'? Are they auditory and express themselves in auditory words, such as, 'I hear what you are saying', 'Your objective comes across loud and clear'? Or are they kinaesthetic and describe things in words we might associate with feelings, such as 'I don't feel happy about your objective', 'I'm all warm about the new plan'? If we mirror the body movements and language of the person or group we are communicating with it gives us better rapport with them. Rapport can be the difference between influencing a stakeholder or losing the sale. If you are working on a sensitive or emotive area of AI, such as healthcare, building rapport with key stakeholders will be fundamental to the success of the project.

1.3.9.1 Dilts' logical levels
The second example is Dilts' logical levels of thinking; we have picked a simple version here to illustrate. It is the fundamental structure in understanding change, demonstrated in Figure 1.2.

Figure 1.2 Dilts' logical levels of change

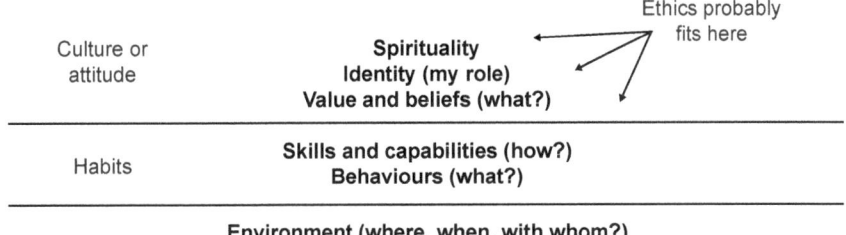

Human logical levels is part of neuro-linguistic programming.[17] This has been a passion and interest of Robert Dilts for decades, and he has made neuro-linguistic programming into a successful international business, helping numerous organisations and individuals.

Our first exploration of the logical levels can be top down or bottom up. So, as an engineer, we might go top down. As an engineer, I value objectivity, science, safety and my values; the things that motivate me are integrity, professionalism, honesty and so on. I believe engineers can make a better sustainable society. The skills I need to do this are mathematics, craft skills, rapport skills and the ability to work in teams. I need the capability to organise my thoughts and draw clear engineering drawings and plans. I behave in a professional manner and work in a variety of places, including a design office and workshop.

This is a pretty basic description of an engineer. How do we make this more explicit? So detailed, in fact, that we could teach other engineers (or a machine) what we are doing so that they are competent and can describe explicitly what they are good at? What happens if we meet other engineers who don't have our values? How does this change their engineering? Neuro-linguistic programming builds models of individuals and organisations so that we can understand them explicitly and are fully aware of what it is that makes us good at something.

At the higher levels of neuro-linguistic programming, spirituality, human identity and values and beliefs play a pivotal role in the success of a congruent outcome. Ethics plays a large role in occupations that involve humans heavily, such as engineering, care, health, politics. These are part of higher levels in Dilts' logical levels. In AI, we need to explore ethics because we are now embracing humans and machines working together. It is possible that machines may have sufficient autonomy that they will influence and determine the outcome of human actions!

The European Union have put together guidelines on AI. They want AI to have an ethical purpose that is trustworthy and technically robust. This, at first inspection, seems easy, but humans are subjective, and measuring what we do is not a simple case of finding an objective measure. Neuro-linguistic programming gives us the tools to work with others and understand those who have different opinions. Simply surrounding ourselves with those we already have rapport with may not necessarily lead to a well-informed outcome. When we think about AI and the need for an ethical purpose, we need to work in the higher levels of Dilts' logical levels. These levels can be emotionally more involved, and we will certainly need tools and skills to do this.

When we frame AI challenges as those that must have a human ethical purpose, the skills we need to make progress shift to human skills. These are far removed from learning computer skills, such as a new parallel programming language or how to use a cloud AI service. If we are not careful, a machine learning approach to AI will not fully embrace humans and machines working together to achieve goals that have this human ethical purpose.

1.3.10 Ethics and trustworthy AI

Ethics is a complex subject even before we start to include the topic of AI, but it has to be a consideration from the outset. We need to consider how ethics should be applied, who is responsible for our AI systems and products, what the law is, how human rights may be affected and even whether we will need to develop robotic rights.

If we ever develop AGI or strong AGI, then robotics becomes a major ethical area to understand.

This book was never intended to be a book on AI ethics, which is a constantly evolving space. This section references what we believe will give you an insight into the subject. From here you can do further in-depth research into the AI ethics that are most applicable to your organisation, your industry or your country.

Some areas of work include:

The Engineering and Physical Sciences Research Council
https://epsrc.ukri.org/research/ourportfolio/themes/engineering/activities/principlesofrobotics/

The Future of Life Institute
https://futureoflife.org/

and

The International Electrotechnical Commission, First International Standards committee for entire AI ecosystem
https://iecetech.org/Technical-Committees/2018-03/First-International-Standards-committee-for-entire-AI-ecosystem

The world of AI is constantly changing, and the rate of change is not linear; you may not have noticed, but it is actually growing exponentially, which basically means it is accelerating faster and faster on an annual basis. Some would argue that its growth is potentially out of control and we need to put the brakes on; others suggest that we just let it run and see where it goes.

Many people have fears that not all AI improvements are necessarily beneficial to humankind or in societies' best interests. For example, there is the potential to weaponise AI, for it to intrude into our lives and privacy by listening and watching everything, and for individuals' identification without consent with covert AI systems. There is also the potential to have control of our personal data taken away from us. So, how should AI be managed and controlled? Or, what happens if we try to do the right thing – whatever that means – and we get it wrong?

1.3.11 General definition of ethics

Ethics is a philosophical subject that manifests itself in individuals as what we term our moral principles; in fact, many people use the words interchangeably. Basically, both ethics and morals relate to 'right' and 'wrong' conduct.

It is worth noting that Ethics, the subject that academics and students study, is treated as singular, for example, 'my elected subject this term is Ethics'. Our moral principles or ethical principles that motivate and guide us are treated as plural. This way we can differentiate between the subject, which is singular, and a person's ethical principles, which are plural.

Further expansion on ethics can be found in *An Intelligent Person's Guide to Ethics* by Lady Warnock.[18]

Morals tend to refer to an individual's own personal principles regarding what is right and wrong. Ethics generally refer to rules provided by an external source or community, for example codes of conduct in the workplace or an agreed code of ethics for a profession such as medicine or law.

Ethics as a field of study is centuries old and centres on questions such as:

- What is a 'good' action?
- What is 'right'?
- What is a 'good' life?

1.3.11.1 The role of AI ethics

We all want beneficial AI systems that we can trust, so the achievement of trustworthy AI draws heavily on the field of ethics. AI ethics could be considered a subfield of applied ethics and technology, and focuses on the ethical issues raised by the design, development, implementation and use of AI technologies.

The goal of AI ethics is therefore to identify how AI can advance or raise concerns about the 'good' life of individuals, whether this be in terms of quality of life, mental autonomy or freedom to live in a democratic society. It concerns itself with issues of diversity and inclusion (with regard to training data and the ends to which AI serves) as well as issues of distributive justice (who will benefit from AI and who will not).

What comes out of consideration of AI ethics are some general themes that most agree on, and these are:

- **Transparency** – we might think of this as how we understand what went wrong when AI gets it wrong. Can we be explicit as to why an AI technology failed? For instance, a NN is generated by an algorithm that we can understand; however, for it to be transparent we would also need to understand explicitly why the NN came to a particular outcome. It gives us a basis on which to learn how the AI system performed and why.
- **Accountability** – we need to be careful or we could get inequality, bias, unemployment and so on.
- **Weaponisation** – creating lethal autonomous weapons is a red line that we must not cross; however, national security is a special subject that is beyond the general use of AI in society.
- **Harm** – the AI must do no harm.

1.3.11.2 What is already in place?

There is already a lot of legislation in place globally and locally, such as the Human Rights Act and data protection legislation that protect our basic human rights. However, often this is abused by the state and by private organisations alike, so why should we trust the creators of AI systems and services to operate responsibly?

1.3.12 Human-centric ethical purpose – fundamental rights, principles and values

A human-centric ethical purpose aims to enhance human capabilities and societal wellbeing. It builds on the Universal Declaration of Human Rights and the EU's Charter of Fundamental Rights of the European Commission. Typical measures of how this might be demonstrated are the United Nations sustainability goals.

There are several bodies and many organisations worldwide currently working on codes of ethics around AI.

1.3.12.1 The European Union
AI ethics guidelines are produced by the European Commission's High-Level Expert Group on Artificial Intelligence.[19] The guidelines are for trustworthy AI.

Trustworthy AI basically has two components:

1. It should respect fundamental rights, applicable regulation and core principles and values, ensuring an 'ethical purpose'.
2. It should be technically robust and reliable, since, even with good intentions, a lack of technological mastery can cause unintentional harm.

It also considers:

- **Rights** – a collection of entitlements that a person may have and which are protected by government and the courts.
- **Values** – ethical ideals or beliefs for which a person has enduring preference and which determine our state of mind and act as a motivator.
- **Principles** – a fundamental well-settled rule of law or standard for good behaviour, or collectively our moral or ethical standards.

At the time of writing, the latest revision of the EU's ethical guidance on AI was produced in March 2019 and we expect it to be revised on a regular basis.

1.3.12.2 Future of Life Institute (US founded)
The Future of Life Institute has developed the Asilomar Principles for AI (https://futureoflife.org/ai-principles/?cn-reloaded=1). There are 23 principles relating to:

>**Research** – goals, funding, policy, cultures, race avoidance (speed of progress).
>
>**Ethics and values** – safety, failure transparency, judicial transparency, responsibility, value alignment, human values, personal privacy, liberty and privacy, shared benefit, shared prosperity, human control, non-subversion, AI arms race.
>
>**Longer term issues** – capability caution, importance, risks, recursive self-improvement, common good.

> The name Asilomar was inspired by the highly influential Asilomar Conference on Recombinant DNA in 1975. The biotechnology community are still influenced by the voluntary guidelines developed at this conference.

1.3.12.3 Individual organisations

Many organisations are in the process of developing their own 'ethics for AI'.

Google is a good example of an organisation that decided to create an in-house AI ethics board, only to hit problems within a week of its launch. Google's Advanced Technology External Advisory Council (ATEAC) was supposed to oversee its work on artificial intelligence and ensure it did not cross any lines, and could be dedicated to 'the responsible development of AI'. It was, however, dissolved after more than 2,000 Google workers signed a petition criticising the company's selection of an anti-LGBT advocate.

1.3.13 Trustworthy AI

What should you be doing? In the same way that we manage health and safety or compliance within our organisations, if we are working with AI or developing AI systems and services then we should adopt specific AI principles and values, for example:

- The principle of beneficence: Do Good.
- The principle of non-maleficence: Do No Harm.
- The principle of autonomy: Preserve Human Agency.
- The principle of justice: Be Fair.
- The principle of explicability: Operate Transparently.

These principles can be found in various disciplines such as medicine, engineering, accountancy and law; the five listed here are taken from *HDSR*.[20] The principles and values need to be applied to our individual use cases, for example autonomous vehicles or personal care systems, or AI to decide insurance rates. Achieving trustworthy AI means that the general and abstract principles documented above need to be mapped into concrete requirements for AI systems and applications.

The 10 requirements documented below have been derived from the rights, principles and values detailed previously. While they are all equally important in different application domains and industries, the specific context needs to be considered for further handling. These requirements for trustworthy AI are given in alphabetical order to stress the equal importance of all requirements.

1. **Accountability** – means to be answerable for actions or decisions, and is essentially about ownership and initiative. Typically, one person should be accountable, although many may be responsible. To hold someone accountable means the person is being asked to explain why they did (or didn't) do something. In our personal lives, we hold people accountable all the time. In AI product development, it means that an employee who is accountable will take responsibility of results and outcomes – they won't presume that this is purely the concern of management.

2. **Data governance** – the process of managing the availability, usability, integrity, confidentiality and security of the data within enterprise systems, and also increasingly across international boundaries. Internally, it is based on data standards and policies that control data usage; internationally, it is based on regulations including Sarbanes–Oxley Act, Basel I, Basel II, Health Insurance Portability and Accountability Act and General Data Protection Regulation, many of which have been enshrined in law (e.g. Data Protection Act 2018 in the UK). Effective data governance ensures that data are consistent and trustworthy and don't get misused; for example, the unauthorised use of personal medical data without the owner's consent for a purpose that was not agreed at the time the data were collected. In an insurance setting, miscollected medical data may have a detrimental effect on a person gaining medical insurance.

3. **Design for all** – effectively universal design, that is, designing a system or service that all can use regardless of gender, age, disability or impairment, rather than the creation of individual systems suited to just part of society. It is effectively design for human diversity, social inclusion and equality.

4. **Governance of AI autonomy (human oversight)** – the framework of guidance and controls that need to be put in place to ensure that AI systems or products do no harm. For example, provision of advice that a suitably qualified person needs to be sitting in the driving seat of an autonomous vehicle while it is in operation; or that a medical diagnostic tool should be not left to decide treatment of a patient without the oversight of a suitably qualified medical practitioner.

5. **Non-discrimination** – the fair and unprejudiced treatment of different categories of people. Typically, it is managed through a non-discrimination policy that ensures equal employment opportunity without discrimination or harassment on the basis of race, colour, religion, sexual orientation, gender identity or expression, age, disability, marital status, citizenship, national origin, genetic information, or any other characteristic.

6. **Respect for (and enhancement of) human autonomy** – autonomy can be defined as the ability of the person to make his or her own decisions, and describes the ability to think, feel, make decisions and act on his or her own.[21] Autonomy includes three facets, consisting of behavioural, emotional and cognitive self-government. In a medical AI context, it is the respect for a patient's personal autonomy to be considered and built into an AI system of service. This is one of many fundamental ethical principles in medicine, and is usually associated with allowing or enabling patients to make their own decisions about which healthcare interventions they will or will not receive.

7. **Respect for privacy** – the ability of an individual, group or organisation to seclude themselves or information about themselves from others. Concerns around privacy often overlap with data security, and may be covered within controls established for that domain. In terms of the growing use of AI within the public environment, it may include the use of surveillance cameras within part of a city making use of facial recognition software without informing people that they are being monitored. Another example may be around mobile phone providers installing geolocation monitoring and data collection apps within software updates.

8. **Robustness** – requires that AI systems be developed with a preventative approach to risks and in a manner such that they reliably behave as intended while

minimising unintentional and unexpected harm and preventing unacceptable harm. An example would be for an AI system to operate as designed within agreed parameters and not learn and apply new behaviours outside those parameters that may harm humans either physically, socially or financially.

9. **Safety** – linked to robustness, safety must become a key element of AI to ensure human acceptance. Rigorous techniques must be established for building safe and trustworthy AI systems and establishing confidence in their behaviour and robustness. This will facilitate their successful adoption in wider society. Ethical design, transparency and testing are key to ensuring safe AI, especially advanced AI systems, to ensure that they are aligned with core human values.

10. **Transparency** – defined in AI terms as being able to explain what is happening within the AI system. How has the AI come to a particular conclusion or decision or action? Without transparency, how can we ensure fairness and remove racial and gender bias, for example? We require increased explicability from AI products and systems, without which trust cannot be established and AI products and systems will not become fully accepted by society.

In 2019, the European Commission's High-Level Expert Group on Artificial Intelligence released *Ethics Guidelines for Trustworthy AI* and consolidated these 10 requirements down to seven:[19]

1. **Human agency and oversight** – including fundamental rights, human agency and human oversight.
2. **Technical robustness and safety** – including resilience to attack and security, fall-back plan and general safety, accuracy, reliability and reproducibility.
3. **Privacy and data governance** – including respect for privacy, quality and integrity of data, and access to data.
4. **Transparency** – including traceability, explainability and communication.
5. **Diversity, non-discrimination and fairness** – including the avoidance of unfair bias, accessibility and universal design, and stakeholder participation.
6. **Societal and environmental wellbeing** – including sustainability and environmental friendliness, social impact, society and democracy.
7. **Accountability** – including auditability, minimisation and reporting of negative impact, trade-offs and redress.

Moving on from principles and requirements, we can now look at how these could be established and implemented. We can use both technical and non-technical methods to achieve trustworthy AI systems that support the principles and requirements mentioned.

Using technical methods, we have:

1. **Ethics and rule of law by design (X-by-design)** – there are numerous ethics boards around the world, some nationally based and some internationally, others corporately or societally based, all establishing various AI ethics and guidelines. There are laws around health and safety, privacy and data protection. These can be incorporated within the design of AI systems rather than bolted on as an afterthought.

For example, 'Privacy by Design' means that organisations need to consider privacy at the initial design stages and throughout the complete development process of new AI products or services that involve processing personal data.

2. **Architectures for trustworthy AI** – establishing trustworthy AI systems architecture, with clear boundaries or limitations of what an AI learning system is capable of, will ensure appropriate controls are introduced to prevent unexpected behaviours or actions that may be detrimental to the user of such a system.
3. **Testing and validating** – the process of evaluating AI software during or at the end of the development process to determine whether it satisfies specified business requirements. Validation and testing also ensure that the AI product actually meets the client's needs.
4. **Traceability and auditability** – traceability is the ability to look at why something occurred and also the effect a request or action has had. Auditability is using traceability to ensure that the requirement or conformity has been met. It can also be used to identify deviations or non-conformance in an AI product or service.
5. **Explanation (XAI research)** – explainable AI is the ability of a human to understand and communicate the results or a solution produced by an AI system. This is compared to a 'black box' solution where even the designers may not be able to interpret why the results were produced.
6. **CE mark** – applicable within the European Economic Area (EEA), a CE mark is an administrative mark for AI and non-AI products which indicates that the manufacturer has checked conformity with health, safety and environmental protection standards. It allows for the free movement of products within the EEA. The CE marking is also found on products sold outside the EEA that have been manufactured to EEA standards.

Non-technical methods include:

1. **Regulation** – the act of controlling an AI product or system either through a law, rule or order. For example, stating that an AI product cannot be sold without a support contract, or an AI service must not be used by minors.
2. **Standardisation** – the AI product or service conforms to a documented standard. For example, all models are produced to the same standard to ensure quality conformance.
3. **Accountability governance** – in AI ethics this translates to answerability, blameworthiness, liability and the expectation of account-giving (i.e. being called to account for one's actions). It is basically where the buck stops.
4. **Code of conduct** – an agreement on rules of behaviour as compared to a code of ethics (which is a set of principles distinguishing right from wrong). A code of conduct normally outlines appropriate actions and regulations for employers, employees or members, as well as what is not acceptable or expected and the legal consequences of breaking the rules.
5. **Education and awareness to foster an ethical mindset** – this plays an important role, both to ensure that knowledge of the potential impact of AI systems is widespread and to make people aware that they can participate in shaping the societal development of such systems.

6. **Stakeholder and social dialogue** – increasingly we are expected to understand the importance of stakeholder engagement and dialogue to help establish social and environmental reporting. We can address some of the key issues surrounding the use of AI and difficulties involved in the implementation and deployment of AI systems by establishing a stakeholder engagement and social dialogue process. For example, in gaining societal agreement to deploy facial recognition and tracking systems, which are contentious, whereas car number plate recognition systems are widely accepted.
7. **Diversity and inclusive design teams** – the creation of teams with diversity of experience, perspective, creativity and inclusion of race, ethnicity, gender, age, sexual identity, ability/disability and location help to create truly inclusive designs.

1.3.14 Trustworthy AI – continual assessment and monitoring

How do we check and confirm that we have hit our AI ethics principals? We do it through ongoing:

- accountability;
- data governance;
- non-discrimination;
- design for all;
- governance of AI autonomy;
- respect for privacy;
- respect for (and enhancement of) human autonomy;
- robustness;
- reliability and reproducibility;
- accuracy through data usage and control;
- fall-back plan;
- safety;
- transparency;
- purpose;
- traceability: method of building the algorithmic system; method of testing the algorithmic system.

So, in summary, here is a very simple mental checklist:

1. Adopt an assessment list for trustworthy AI.
2. Adapt an assessment list to your specific use case.
3. Remember that trustworthy AI is not about ticking boxes, it is about improved outcomes through the entire life cycle of the AI system.

1.4 SUSTAINABLE AI

As well as trustworthy and robust AI we need to ensure that what we are doing in AI is sustainable. In 2015, the United Nations produced its Agenda for Sustainable Development and identified 17 sustainable development goals (SDGs) and 169 targets. Amazingly, it was agreed by 193 countries, and had effectively produced 'The framework for good'.

An open paper produced by a number of sustainability experts from around the world in 2019 identified that AI may support the achievement of 128 targets (76 per cent) across all SDGs, but AI may also negatively impact or inhibit 58 (34 per cent) of those targets.[22] The paper further broke down the targets within the three pillars of sustainable development: society, economy and environment.

We should recognise that AI has a role to play in sustainability and that the current capabilities of AI, including automating routine and repetitive tasks, analysing Big Data and bringing intelligence and learning to various processes, have expanded and continue to expand our capacity to understand and solve complex, dynamic and interconnected global challenges such as the SDGs.

At the time of writing, with just 10 years remaining (up to 2030) to achieve the ambitions outlined in the United Nations SDGs, we should use AI to achieve those SDG targets we know it can contribute to, while at the same time identifying the impact and inhibitions that AI will have and mitigating the negative effects where possible. AI may also trigger inequalities that could act as inhibitors on some of the SDGs. For example, in helping to deliver the achievement of the 'No Poverty' SDG, AI can help to identify areas and pockets of poverty and their causes; however, it may also increase the gap between poor and rich countries and between the poor and rich within each country.

Guided by the United Nations SDGs, it is now up to all of us – businesses, governments, academia, multilateral institutions, non-governmental organisations (NGOs) and others – to work together to accelerate and scale the development and use of AI in the interest of achieving the SDG targets, while at the same time recognising possible negative effects and addressing them.

We also need to consider that not all stakeholders in the development and implementation of AI will be interested in sustainability, and may well have their own measures, such as return on investment.

> Rogue actors may develop malicious AI. We must be on guard and develop intelligent cyber protection. This can be considered as the development of robust trustworthy AI.

1.4.1 Political, Economic, Sociological, Technological, Legal and Environmental (PESTLE)

Although the pursuit of SDGs may be seen as altruistic, that does not say that it should be left to others to pursue. When an organisation is designing, developing or implementing an AI solution, they should consider at least undertaking a PESTLE analysis. This is a framework to analyse the key factors influencing an organisation from the outside.

PESTLE is a strategic and systematic way of learning about key factors that influence a range of stakeholders. It can be used by an organisation for tactical decision making and often is.

1.5 MACHINE LEARNING – A SIGNIFICANT CONTRIBUTION TO THE GROWTH OF ARTIFICIAL INTELLIGENCE

In this section, we need to make the distinction between machine learning and artificial intelligence. This is no easy task until we remind ourselves that AI is a universal subject that can help any pursuit of learning from experience to achieve a goal.[9] ML is only part of this – the distinction starts with the definitions of an AI agent and ML. Tom Mitchell's definition is the one usually quoted;[23] it relates well to digital computing that is part of all our lives.

AI is about intelligent entities, or AI agents, interacting with an environment to achieve a goal. Not just one entity but multiple. AI is about humans and machines, how they interact, how they learn, what they experience. AI asks us to be explicit about humans achieving their goals. Can we build machines that we can work with to improve us as humans? AI asks hard questions of consciousness, philosophy, ethics, science. It deals with complex phenomena that we do not currently have the machines to explore. AI of the future is about how humans and machines will co-exist. In 2019 Stuart Russell released his latest book on human compatible AI, *Human Compatible: Artificial Intelligence and the Problem of Control*,[1] which captures the true nature of AI and how, used wisely, we can benefit from a future of humans and machines.

As we are coming to realise the benefit of digital computing ML, this will unlock the potential of AI in other areas:

- engineering and building intelligent entities;
- medicine and improving health- and social care;
- business analytics, and others.

When we simply think about products, representing the world in a digital simulation, or what is (at the moment) ones and zeros, it is not ideal. Digital computation has its limitations, and we need better machines. We have run out of digital processing power. AI is much more than high-performance computing and programming of machines that deal with ones and zeros – digitally simulated pizza and weather don't taste of anything and do not get us wet.

The AI machines of the future will incorporate digital computers, but, when we think about it, it's actually hard to represent the mathematical operations we need. We are limited by the processing power, energy and accuracy of today's technology. And a result, we can only concentrate on narrow ML, focused on specific well-defined tasks or goals. These tasks are defined by Tom Mitchell's often quoted definition (p. 2):[23]

> A computer program is said to learn from experience, *E*, with respect to some class of tasks, *T*, and performance measure, *P*, if its performance at tasks in, *T*, as measured by, *P*, improves with experience, *E*.

The examples given of these types of tasks are playing games such as chess, checkers and draughts. Modern-day games include simulation games and very advanced strategic, well-engineered games like Go; these types of games can be explicitly defined on a digital computer. Practical examples in the real world include optimising where aircraft park at an airport or the logistics of delivering a parcel; again, a reasonably well-ordered and engineered environment in which ML can optimise something.

ML is focused on explicitly defining a problem that can be solved on a computer. These problems can be complicated, non-linear and statistical. In simple terms, if we are to use ML in our AI, we must be able to represent our problem mathematically and in such a way that it can be solved by a machine. Today, we typically use digital computers. However, quantum, analogue, optical and biological computers are on their way.

Digital ML has become very popular recently with the success of convolutional deep neural networks. These numerical techniques have given us an understanding of how the human mind solves problems. So, when we think about ML, we can think about an AI agent learning from data. These machines are now so good at playing games that they can beat the world champion at these types of games.

The AI agent is much more than a narrowly focused computer program. ML works on data in the computer. We must work really hard to think of this as an interface with actuators and sensors in an environment. It is sensible to think, here, that ML learns from data in a computational environment. This is a good starting point to opening up the world of AI. AI is about humans and machines working together to achieve goals. We might even go on to say that ML is an AI enabler setting the foundation for a future of humans and machines.

1.6 SUMMARY

This chapter has been a whistle-stop tour of human and AI intelligence. It has drawn out key concepts such as agents, ethics and machine learning and their historical context. It has highlighted that humans are subjective and objective conscious beings, something AI is nowhere near achieving. With our introductory knowledge of human intelligence and the progress of AI over the past few centuries, we can look in more detail at ethical AI for human good and how this may evolve into products and services in an age of humans and machines.

2 ARTIFICIAL INTELLIGENCE AND ROBOTICS

In this chapter, we explore AI in terms of agents and robotics. While doing so, we should remind ourselves of the definition of human intelligence discussed in Section 1.1, from the *Encyclopedia Britannica*:[8]

> Human intelligence: mental quality that consists of the abilities to learn from experience, adapt to new situations, understand and handle abstract concepts, and use knowledge to manipulate one's environment.

It is really helpful to use the AI agent as the schematic of the artificial human, and it has a simple analogy in that it is similar to a robot. Unfortunately, this simple analogy has not been used by roboticists. Their focus hasn't been human intelligence or the framework for the academic study of intelligence. Instead, they have been focused on the engineering of a robot, which is complex and sophisticated engineering. We are now seeing these disciplines collaborating and exciting new areas emerging. It's an exciting time because advances in ML have unlocked the ability of computers to teach themselves how to achieve a goal, in either a digital simulated environment or a practical engineered environment. It opens up a world of opportunities: autonomous smart cities, smart roads, smart homes, smart TVs, smartphones, smart watches, smart nano-bots.

We can think of AI products, big or small, as AI agents. They can have a perception of the environment they are in and then act on the environment to achieve a goal. ML gives us the digital examples of how this can be achieved. It can also be used to give the AI agent or products autonomy to learn from experience, for example how to undertake a task or tasks and achieve a goal. The functionality we require for intelligent agents links simply with machine learning. Games can be used to understand the interaction of autonomous agents.

Agent-based modelling is a fundamental method useful in academia, engineering, business, medicine and elsewhere. Through this, we can understand the emergent behaviour of multiple agents operating in an environment.

2.1 UNDERSTANDING THE AI INTELLIGENT AGENT

We start with the academic description of an AI agent from Russell and Norvig.[9] They define a rational agent as:

> For each possible percept sequence, a rational agent should select an action that is expected to maximize its performance measure, given the evidence provided by the percept sequence and whatever built in knowledge the agent has.

This can be illustrated as a learning agent (see Figure 2.1). The AI agent is the schematic of how AI describes a human acting in an environment; the traditional engineering concept of a robot or control system is the dashed line. ML has certainly been used inside this dashed line for many years – we will introduce Perceptrons in Chapter 4. With the advent of deep learning and its more recent successes, this dashed line is now moving to encapsulate the whole AI agent. Learning from experience now means the potential for autonomy. We humans sometimes call this 'agency'.

> Agency comes from late middle English (someone or something that produces an effect), founded on the Latin word *agere*, 'to do/act'.

Figure 2.1 The learning agent – based on the structure of agents in Russell and Norvig[9]

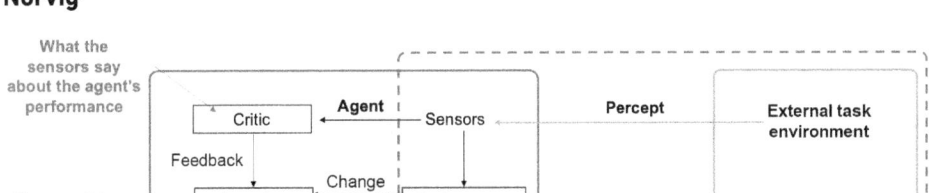

Humans revere agency, and humans are agents that act in an environment.[24] Humans want the free will to act – it is important to us. For a robot to have free will to act, we call this autonomy. Autonomous robots are usually portrayed as baddies in movies because we try to wrap meaning around a robot's actions. Robots, at least for now, are objective and rational and do what we tell them to do (sometimes this may have bad consequences). Humans are subjective, our free will is subjective and when we think about agency, we must consider the subjective side of our actions – this leads to the philosophical and moral understanding of humans. In defining an AI agent, we rely on the agent being rational, in the same way that engineers use objective science to design robots.

2.1.1 The four rational agent dependencies

We expect AI agents to be rational, and this depends on four things:[9]

1. The performance measure that defines the criterion of success.
2. The agent's prior knowledge of the environment.

3. The actions that the agent can perform.
4. The agent's percept sequence to date.

These four elements are required to provide a fundamental academic basis for a rational agent. If we think about it, we need all four so that others can reproduce our results. Without them the agent would be ill-defined.

We know that AI is an academic subject based on the scientific method. Rationality helps when we are thinking about AI agents. So, in designing an AI agent – maybe a robot or a data robot or a simulated agent looking at how humans behave – we expect them to act rationally and do what we ask them to do. Going back to Russell and Norvig's definition of the rational agent:[9]

> For each possible percept sequence, a rational agent should select an action that is expected to maximize its performance measure, given the evidence provided by the percept sequence and whatever built in knowledge the agent has.

We can see that when we design an AI system, a robot or an engineered product, it is telling us that it must perform as we design it to perform.

2.1.2 Agents – performance measure, environment, actuators and sensors

If we are an engineer asked to design an intelligent entity, then it makes a lot of sense to define a performance measure by which its performance can be measured, define the environment in which it operates and the actuators and sensors it uses to manipulate its environment. By doing so, we create an intelligent entity that can learn from its experience. If we take a humanoid robot builder as an example, then we could describe each as the following:

- **Performance measure** – speed to lay a fixed number of bricks in a wall.
- **Environment** – the building area or room.
- **Sensors** – tactile sensors, power sensors, visual and sound sensors and so on.
- **Actuators** – hands, body, feet, legs and so on.

Somehow, this seems easy, but it is not explicit mathematically. We could think about building a digital twin so we can model it and learn. By this we mean we can build a simulation environment for the robot to learn from, a bit like a computer game. Just as we teach pilots how to fly in a flight simulator, we can teach a robot in the same way. Or perhaps we can let the robot use reinforcement learning to teach itself. For a more detailed understanding of these concepts mathematically, see Russell and Norvig.[9]

An agent might want its own map of the environment. We call this the state of the world of the agent. It is the agent's internal world that they might use to make up for a lack of information. This world can be simple, for example a length of railway track and the temperature of the track might be all we need to control our autonomous train. This is the atomic state. It might be a little more involved where we use a range of sensors to determine the state of the world or environment we are operating, for example rain, temperature, wind and humidity determine the weather that our train is travelling in.

This is a factored state. Or, we could use many different objects and the relationship of those objects to determine the state of our world. This is a structured state. The objects and the relationship of those objects give the agent a more detailed understanding of the environment it is operating in. It might use AI to predict the outcome of different actions; this is shown schematically in Figure 2.2.

Figure 2.2 The state of the world – the agent's internal world that can make up for incomplete information or help to make decisions

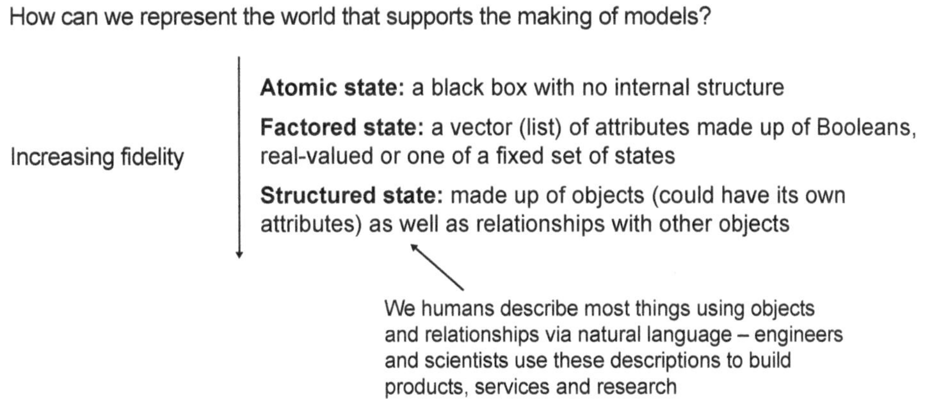

How can we represent the world that supports the making of models?

Increasing fidelity

Atomic state: a black box with no internal structure

Factored state: a vector (list) of attributes made up of Booleans, real-valued or one of a fixed set of states

Structured state: made up of objects (could have its own attributes) as well as relationships with other objects

We humans describe most things using objects and relationships via natural language – engineers and scientists use these descriptions to build products, services and research

---

In understanding the agent, we need to understand the mathematics of its internal world, and so the use of a digital twin seems a sensible idea if we need to train an intelligent entity. What happens if we need multiple intelligent agents working together to achieve a goal?

2.1.2.1 Agent-based modelling (ABM)
ABM looks at multiple agents working together to achieve a goal.[25] We might think of this as how bees build a wonderful structure to produce honey. An agent-based model can look at how the bees communicate, how they maintain order or control; also, how an intelligent entity would work around humans or other animals. ABM is a really useful approach to understanding emergent behaviour. When we think about ABM, controlling multiple agents or finding a hierarchical controls system like subsumption will become a key concept.[26] This is a growing area of interest and we touch on ABM later in Section 2.3.1 on intelligent robotics.

2.1.3 Types of agent: reflex, model-based reflex, goal-based and utility-based

In AI, an intelligent agent refers to an autonomous entity that acts, directing its activity towards achieving goals in an environment using observation through sensors and consequent actuators, and is therefore intelligent. There are basically four rational types of agent, examples ranging from a relatively simple heating system control agent, building up in complexity to an automated medical general practitioner (GP/doctor) agent.

Let's look at each of them.

2.1.3.1 Reflex agent
In AI, a simple reflex agent is a type of intelligent agent that performs actions based solely on the current situation. The agent does this through predetermined rules for these conditions. This is commonly referred to as the condition–action rule, the program selects actions based on the current percept. It's simple to understand and program.

Example: central heating overheats, program selects the action to switch off the power.

2.1.3.2 Model-based reflex agent
In a model-based reflex agent, the agent has a model of the world that it can call upon if needed:

- It can make up for a lack of sensors by using a virtual world that doesn't need sensor data.
- The program now looks at the percept and updates its own internal world (state).
- The program can then assess the possible actions and future states, and uses the reflex agent approach to determine what actions to take.

Example: an underwater vehicle loses visual sensor data by stirring up silt – there's not enough light, but it can continue using its 3D mapped geometrical model.

2.1.3.3 Goal-based agent
Goal-based agents can further expand on the capabilities of the model-based reflex agents, by using 'goal' information. The goal information describes situations that are desirable. Search and planning are the subfields of AI devoted to finding action sequences that achieve the agent's goals:

- The program needs more than just sensors and an internal world to implement the agent's functionality: it needs a goal.
- These programs are more versatile and flexible and can adapt to changes.

Example: an autonomous vehicle can choose one of five exits from a motorway, goals could be the safest, quickest, most scenic, cheapest and/or shortest route.

2.1.3.4 Utility-based reflex agent
A utility-based agent is an agent that is a step above the goal-based agent. Its actions are based not only on what the goal is, but the best way to reach that goal. In short, it is the usefulness or utility of the agent that makes it distinct from its counterparts. The utility-based agent makes its decisions based on which action is the most pleasing or the most effective at achieving a goal. The goal is measured in terms of the agent's utility.

> Utility is the scientific way economists measure an agent's happiness; it is more versatile than a simple binary yes/no or happy/unhappy output, it actually measures how useful it is.

Example: an AI chatbot doctor assesses the efficacy of a treatment based on a patient's individual medical history and current health needs – that is, what utility is this medication to the patient?

> The 'learning agent' is an overarching schematic of an AI agent that can learn how to do something. This specialised agent is capable of learning from its experiences, unlike the four intelligent agents detailed above that act on information provided. Typically, it starts with some basic knowledge and adapts through learning. A learning agent is able to:
>
> - perform tasks;
> - analyse its own performance;
> - look for new ways to improve on those tasks.
>
> Moreover, it can do this all on its own.
>
> An example of a learning agent is a personal assistant like Siri or Alexa, which can look up your local weather conditions on the internet and advise you whether to wear a coat or not because it has learned that you like to go for a walk when you haven't slept well.

2.1.4 The relationship of AI agents with machine learning

An agent should be autonomous. What this means for an intelligent entity is that it can learn, and make up for incomplete knowledge and incorrect prior knowledge. For example, the Deep Blue machine that beat the world chess champion had not learned how to do that by itself. It relied on the prior knowledge of the programmers to win its games.

A truly rational agent should be autonomous – it should learn to make up for incomplete and incorrect prior knowledge – but I think it's safe to say that we are concerned about agents having too much autonomy! It should be recognised that, at present, a robot can't just wake up one day and start doing something; we need to be realistic and give the robot a sensible starting point.

An agent lacks autonomy if it relies on prior knowledge from its designer. But there needs to be some prior knowledge to get it going. It will certainly need an initial condition.

2.2 WHAT IS A ROBOT?

For those of you that didn't read the brief history of AI in Section 1.3, the word 'robot' came into use in Czech-born Karel Čapek's 1920 stage play *R.U.R.*, which stands for *Rossumovi Univerzální Roboti* (Rossum's Universal Robots). The play opens in a factory that makes artificial people called 'robots'. The robots in the play are closer to the modern idea of androids or even clones, as they can be mistaken for humans and can think for themselves. Without giving the plot away too much, after initially being happy working for humans, a hostile robot rebellion then ensues that leads to the extinction of the human race. As we mentioned earlier, the word 'robot' comes from the Slavic language word *robota*, meaning 'forced labourer'. Nowadays we consider a robot to be a machine that can carry out a complex series of tasks automatically, either with or without intelligence.

Robots have been around for over 60 years. In 1954 George Devol invented the first digitally operated and programmable robot, called the Unimate. In 1956 he and his partner, Joseph Engelberger, formed the world's first robot company, and the Unimate, the first industrial robot, went online in the General Motors automobile factory in New Jersey, USA, in 1961. Robot development has been ongoing around the world ever since, and robots in various guises are now in general use in most factories and distribution centres around the world, quite often working with and alongside humans on complex production lines.

You may even have allowed domestic robots to invade your own home – robotic vacuum cleaners, first appearing in 1996, are the most widely recognised. The Roomba, first introduced in 2002, is an autonomous robotic floor vacuum cleaner with intelligent programming and, despite nearly 20 years of development, still struggles with obstacles such as dog faeces, cables and shoes. Similar technologies are often employed to mow lawns, but also struggle with discarded garden toys and, of course, humans' best friend.

One of the most widely recognised robots is ASIMO (Advanced Step in Innovative Mobility), which is a small child-like humanoid robot created by Honda in 2000 (see Figure 2.3). It has been developed continually since then and now has become one of the world's most advanced social robots. Unfortunately, ASIMO has recently gone into retirement, but Honda say that the technology will be diverted into 'nursing care' robots in the future.

Although many people look towards the positive aspects of robot development, others see robots in a negative light and project a future in which robots take jobs from human workers. However, 'crystal ball gazers' generally believe that the AI and robotics industries will probably create as many new jobs as are lost through increased automation.

Although we have focused so far on intelligent robots, the vast majority in operation are unintelligent (i.e. there is no learning) and simply follow a computer program to carry out a limited action. Now, if we apply AI techniques to robots and allow them to learn and work autonomously across a number of areas, we could equally have created either a useful home assistive robot or the robot from *The Terminator*, depending on your point of view.

2.2.1 Robotic paradigms

A paradigm is a philosophy or set of assumptions and/or techniques that characterise an approach to a class of problems. The word paradigm comes up a lot in the academic, scientific and business worlds. When you change paradigms, you are changing how you think about something; it is a way of looking at the world – it is also a set of tools for handling and solving robotic problems.

ARTIFICIAL INTELLIGENCE AND ROBOTICS

Figure 2.3 Honda's ASIMO, conducting pose captured on 14 April 2008 (reproduced from https://commons.wikimedia.org/wiki/File:ASIMO_Conducting_Pose_on_4.14.2008.jpg under Wikimedia Commons licence)

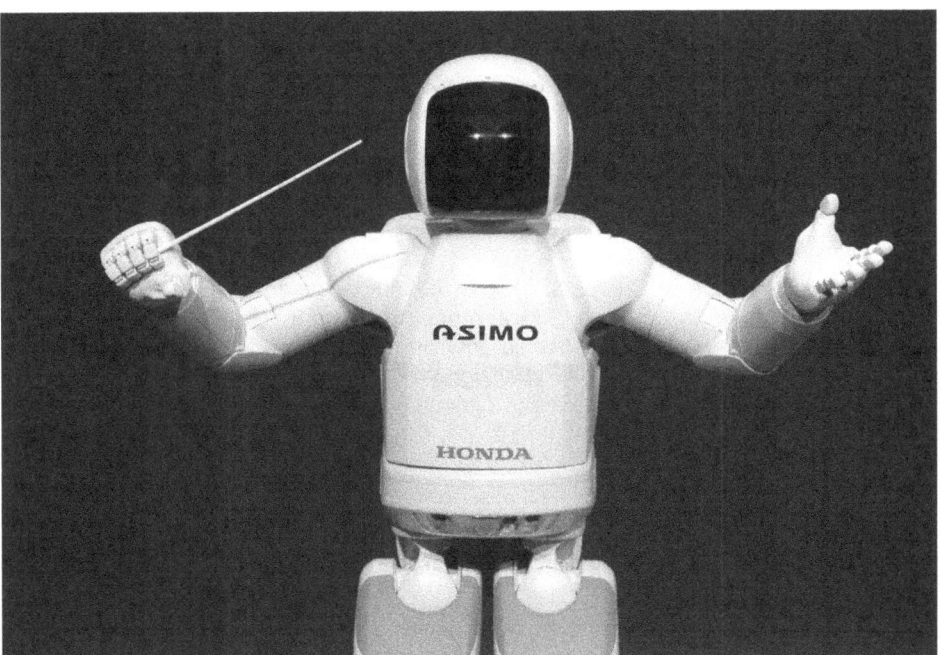

Specifically, in robotics, a robotic paradigm is a mental model of how a robot operates. A robotic paradigm can be described by the relationship between the three primitives of robotics: sense, plan and act. It can also be described by how sensory data are processed and distributed through the system, and where decisions are made.

In robotics there are three paradigms:

1. hierarchical;
2. reactive;
3. hybrid hierarchical–reactive.

These three paradigms are used for organising intelligence in robots – and selecting the right paradigm makes problem solving easier. The hierarchical, reactive and hybrid hierarchical–reactive paradigms all use two or three robotic primitives:

- sense (uses information from sensors and the output is sensed information);
- plan (sensed or cognitive information and the output is a directive);
- act (sensed information or directives and the output are actuator commands).

In robotics there are two ways to describe a paradigm: either by the relationship of sense, plan and act:

- Functions the robot undertakes are categorised into the three primitives.

Or by the way sensor data are utilised and organised:

- How is the robot or system influenced by what it senses?
- Does the robot take in all sensor data and process them, for example, or are sensor data processed locally?

2.2.1.1 Hierarchical

Here there is a fixed order of events, which is based on top-down planning (see Figure 2.4). These robots tend to have a central model to explicitly plan from. It tends to be highly controlling in a highly controlled environment and difficult to use in complex environments. This is probably OK in a highly organised and engineered world, but it does ignore the human cognitive and biological elements, for example we can act and sense at the same time. An example is a robot that selects an item from a fixed location.

Figure 2.4 Hierarchical paradigm in robotic design

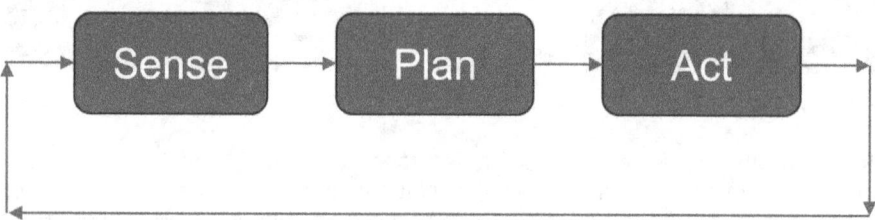

Hierarchical
(1967 onwards)

2.2.1.2 Reactive

Reactive robots are always sensing and acting. As the sensing changes so its behaviour changes (see Figure 2.5). Couple these behaviours together and you are not stuck with one sense action, which saves planning time. To gain a more intuitive understanding, it's like the knee jerk reaction to being surprised – if we place our hand on a hot surface we pull away really quickly, and that is automatic. With the reactive paradigm, the robot reacts to a sense input. An example is a robot that maintains a set direction, but if a sensor senses an obstacle the wheels react and turn away from the obstacle.

Figure 2.5 Reactive paradigm in robotic design

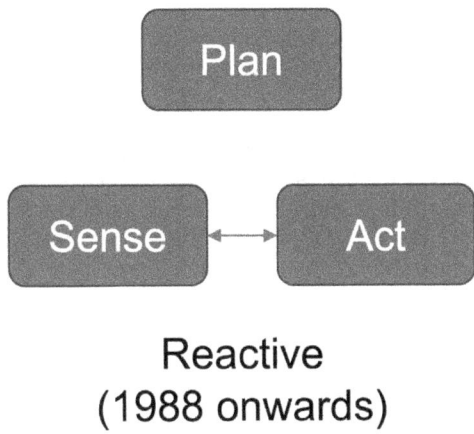

Reactive
(1988 onwards)

2.2.1.3 Hybrid
This paradigm is a combination of the hierarchical and reactive paradigms. These robots plan their initial action to achieve a goal. Sense and act are then informed by a plan, (see Figure 2.6).

Figure 2.6 Hybrid deliberative/reactive paradigm in robotic design

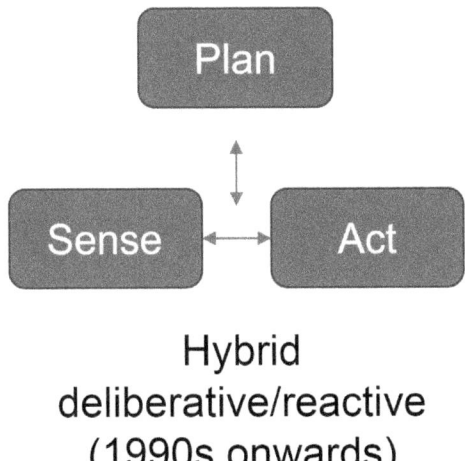

Hybrid
deliberative/reactive
(1990s onwards)

Sensors can update the plan (e.g. identify an object), and this ability allows each of the sensors, actions and planning to run at a speed that is appropriate. This paradigm is a natural evolution of robotic design that allows different parts of the robot to be controlled by something that is appropriate for what it is doing. Some of our cars have automatic braking systems that help us brake more effectively if there is an obstacle or an icy road. These systems are reactive, but we can think of the driver as the hybrid deliberative/reactive part of the system – the driver does the sense, plan, act. An engine management system can be thought of as a 'Sense – Plan, Act' robotic paradigm: it takes multiple readings from the engine (temperature, engine load, air flow rate), plans the optimum fuel to air mixture and ensures the car delivers the maximum power with the least emissions.

2.2.1.4 Russell and Norvig – a general learning agent

This information together reminds us of which part of Russell and Norvig's learning agent schematic is typically associated with a robot (see Figure 2.1). Note the learning element on the left-hand side – autonomy would allow the robot to learn from experience, from its environment, and make up for unknown or missing data.

2.3 WHAT IS AN INTELLIGENT ROBOT?

This is possibly the shortest section on our topic of robots. An intelligent robot is one that uses AI or, as Robin Murphy defines it,[27] 'an intelligent robot is a mechanical creature which can function autonomously'.

> We must note that the idea of a robot here is a mechanical robot that we see in industry or a humanoid robot that could be our personal assistant. Robots can also be found in the data world of digital computers. Here, robotic process automation (RPA) is a term used to describe agents that work with data in an enterprise setting. RPA isn't usually an intelligent agent, even though the rules it follows are complex.

2.3.1 Intelligent robotics to intelligent agents

The advent of the intelligent learning agent plus reinforcement learning neural networks making significant progress over the past few decades have laid the foundation for the concept of autonomous robots to become a reality. Autonomous vehicles are commonplace throughout the United States now; many states allow testing of robotic vehicles in a driverless setting. Ideas such as ABM, digital twins and simulation are maturing into fertile ground for development – which really means learning from experience. Complete ecosystems for robotic development are closely linked with learning environments for AI agents. ABM involving humans and machines seems the next natural step – multiple agents, multiple intelligent robots and, of course, multiple intelligent humans interacting!

2.4 SUMMARY

In this chapter we have built up our understanding of the intelligent agent. This is a schematic of how AI studies and interprets human intelligence, in particular how learning from experience is encapsulated in the learning agent and how this can be related to a robot. The idea that a robot can learn from experience leads us to autonomous robots. The next chapter highlights the benefits and challenges of AI moving forward.

3 APPLYING THE BENEFITS OF AI AND IDENTIFYING CHALLENGES AND RISKS

There must be clear benefits to the creation and adoption of AI technologies, otherwise why would we bother to continue along this path? However, along with benefits will come certain challenges and risks, which need to be identified, recognised and addressed.

3.1 SUSTAINABILITY – HOW OUR VALUES WILL CHANGE HUMANS, SOCIETY AND ORGANISATIONS

We have previously talked about how AI will affect us as individuals in our workplace by improving efficiencies and augmenting what we do and are capable of. It will also affect us in our everyday lives, removing inequalities and allowing us to become more human with a human-centric ethical purpose, undertaking tasks that involve creativity and empathy, among others.

How will AI change our society? We need to consider what implications AI will have on wider society now and in the long term, and what freedoms and human values we are prepared to give up in return for the benefits we will enjoy. Increased use of AI and wider adoption will come with some risks and challenges, such as the potential of its weaponisation, but it may also solve climate change and world poverty. How can we balance the risks versus the benefits?

As a society, we need to start preparing now for how we manage AI today and in the future. For example, would you be prepared to give up your job to an AI entity but in return receive a basic universal income? Maybe, but would you be prepared to give up your child's future career dreams and aspirations for short-term gains here and now? What happens if we charge off with the aim of utopia and get it wrong?

3.1.1 Intergenerational equity

Intergenerational equity is the concept or idea of fairness or justice in relationships between generations. It is often considered as a value concept that focuses on the rights of future generations. It describes how each future generation has the right to inherit the same level of diversity in natural and cultural resources enjoyed by previous generations and equitable access to the use and benefits of these resources. A common example often quoted is current generations running up debt that will have to be paid off by future generations to come.

The study of intergenerational equity and view of sustainability is based on three pillars: economic, social and environmental (see Figure 3.1).

Economic – many economists have tried to predict the effect that AI will have on the economy. PwC forecast in 2017 that AI could contribute up to US$15.7 trillion to the global economy in 2030,[28] more than the current output of China and India combined. Of this, US$6.6 trillion is likely to come from increased productivity and US$9.1 trillion is likely to come from consumption-side effects.

Social – in the UK during the 1970s, whole communities felt the impact of large organisations closing down operations, including dockyards, mines and airlines. AI brings with it big headlines of major disruption and potential job losses. In an era of humans and machines, the possibility exists for humans to move onto higher value work or to enrich and develop their talents while machines take away the heavy lifting, reducing the burden on human effort. Sustainability is part of the EU's AI guidelines, in particular the human-centric ethical purpose. It is essential, therefore, that assessing the impact of AI on society is a part of that.

Environmental – the AI carbon footprint most certainly has an environmental impact, which needs to be factored into its business case. Donna Lu suggested in a 2019 article in *New Scientist* that 'Creating an AI can be five times worse for the planet than a car.'[29]

Figure 3.1 The three pillars of sustainability

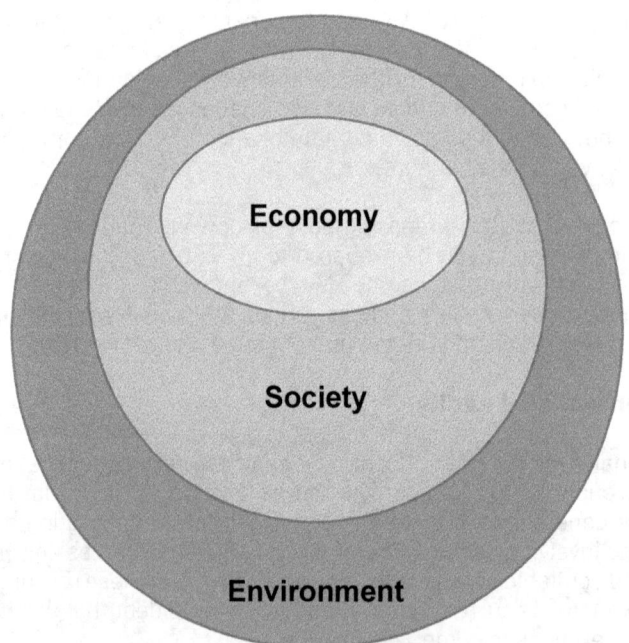

The emergence of AI and its progressively wider impact on many sectors across society requires an assessment of its effect on sustainable development. We must address concerns such as job displacement and social manipulation, design and develop AI technology safely and inclusively, and ensure equitable access to its benefits.

> Social manipulation can come in many forms. A simple example is making bank accounts only cheaper for those who have the ability to use a mobile application, or only giving humans access to AI healthcare if they have mobile applications.

3.1.2 Design for sustainability

Most organisations already have policies around sustainability. AI can help to formulate sustainable processes to underpin our policies but also has to be factored into our sustainability approach. Key considerations should include:

- **Be intentional about sustainability** – use AI to drive organisational sustainability goals and help transition the organisation to a new 'normal' state rather than arrive there by accident.
- **Partner with employees** – engage with employees to identify repetitive processes and activities that can be replaced by AI systems, freeing them up to do higher quality work. This may seem obvious, so positive outcomes for employees need to be demonstrated and communicated.
- **Water and electric conservation** – development of smart buildings using AI-driven technologies will allow organisations to reduce their carbon footprint and save money at the same time.
- **Supply chains** – use AI to manage supply chains that eliminate unnecessary waste; these are known as intelligent integrated supply chains. Integrated supply chains mean we organise and coordinate more closely how we deliver something (a product or service).
- **Develop a recycling programme** – use AI to help reduce, reuse, recycle and repurpose products and goods.
- **Chemical management** – use AI to enable the safe management of harmful chemicals used in manufacturing and through product disposal, for example lithium batteries.
- **Purchase only energy efficient products** – use AI to identify the most energy efficient products within procurement.
- **Develop sustainability work policies** – sustainable work policies (e.g. working from home and using AI) can act as both a driver and an enabler of more effective work.

3.2 THE BENEFITS OF ARTIFICIAL INTELLIGENCE

The are many benefits of AI. Social media frenzy, television programmes and movies all extol the success of possible futures of AI, but we need to be practical because chasing ivory tower academic dreams won't bring to life the benefits. As we move forward, we need to be explicit and careful about what AI and ML can achieve.

Paul R. Daugherty and H. James Wilson have written about their experience doing just that as management consultants: *Human + Machine: Reimagining Work in the Age of AI.*[30] In their book they observe how AI is applied in many projects. They describe human-only systems, machine-only systems and human plus machine systems. They paint a picture of how the future may look, what roles humans and machines will play and what humans and machines will do together.

3.2.1 Human-only systems

What do humans do well?

- leadership and goals;
- creativity;
- empathy;
- judgement.

In Chapter 1 we marvelled at what the human being is capable of. Machines are some way off this – decades or perhaps centuries off. Humans are good at the subjective viewpoint. We are creative; for instance, only a human, in this case Einstein, could have imagined a thought experiment of travelling on a light beam. It changed our understanding and the world as we know it today.

So, humans will be the creative powerhouse. We will let machines do the heavy lifting, leaving us to do the higher value work. Ambiguity is also a role for humans – what happens if the AI is stuck because of a contradiction? It is up to a human to step in and determine the outcome. We can think of many situations where human intervention might be needed (e.g. a confused dementia patient unable to answer questions, or an ethical 50/50 contradiction). A human who is confused could easily confuse an AI system; other humans need to be on hand to support that person.

Humans must also provide the leadership; we must set the goals and the ethical standards. These will form principles, values and rights. We will rely on these in law and, where the judgement is subjective and ambiguous, it will be humans who do this high value work.

3.2.2 Machine-only systems

What do machines do well?

- monotonous tasks;
- prediction;
- iteration;
- adaption.

These four points describe what machines do well, and this is easy to see with digital computers. We use them to do monotonous tasks such as lots of accurate calculations. Predicting and iterating are useful in digital computers for the solution of our ML problems – lots of calculations, repetitive iterations, finding a solution. Adaption is a trait of future machines, when machines will adapt to the needs of the day. Machines in this sense will also adapt to extreme environments; these could be deep in the ocean or on other planets where humans are exposed to extremes. More practically, preparing food in sterile conditions or operating on humans in an operating theatre could be better done by machines.

3.2.2.1 Narrow (weak) artificial intelligence, artificial general intelligence and strong artificial intelligence

Weak or narrow AI is AI focused on a specific task.[9] Popular ML of today, the capability we find on cloud digital services, is narrow AI, or narrow ML, focused on a specific task (e.g. supervised and unsupervised learning). Examples of narrow ML are support vector machines, decision trees, k-Nearest Neighbour algorithms. They test hypothesis based on a specific task. The learning from experience is on a specific focused task.

AGI is a machine capable of learning any intellectual task that a human can. It is hypothetical and not required for AI and ML in general. This hypothetical type of AI is often the subject of speculation and sometimes feared by humans. We are nowhere near AGI, and it could be decades or even centuries before we are close to achieving it.

AGI can be taken one step further. Science fiction is fond of machines that are assumed to have consciousness; strong AI is AGI that is also conscious. Consciousness is complex and difficult. Some consider it to be the hardest problem in AI. This is covered in more detail in Chapter 1, particularly Section 1.3.8.

AGI and strong, or conscious, AGI is not currently feasible and is not realistic in the foreseeable future. It is an active area of research, but not a requirement for AI or ML.

> The origins of consciousness are not known, we can only speculate as to what consciousness is. Professor David Chalmers has posed two questions of consciousness, the easy question and the hard question:[31]
>
> 1. The easy question is to explain the ability to discriminate, react to stimuli, integrate information, report mental states – and we can study these with the scientific method.
> 2. The hard question is about how and why physical processes give rise to experience. Why do these processes take place 'in the dark', without any accompanying state of experience?
>
> Essentially part of being human is a subjective experience. Understanding this conscious experience is not going to be easy. Professor Chalmers and Professor Roger Penrose both think that the answer to the hard question could be as hard as quantum mechanics. Professor Penrose has proposed a mechanism or foundation for consciousness based on quantum mechanics. More detail of these concepts can be found in the reading list at the end of this book.[32]

There has been considerable debate about consciousness. It is an open question and an academic exercise. One of the core complete rejections of strong AI, at least in digital computers, is presented by Professor John Searle. He argues using the Chinese Room thought experiment that strong AI cannot occur, and states:[33]

> Computational models of consciousness are not sufficient by themselves for consciousness. The computational model for consciousness stands to consciousness in the same way the computational model of anything stands to the domain being modelled. Nobody supposes that the computational model of rainstorms in London will leave us all wet. But they make the mistake of supposing that the computational model of consciousness is somehow conscious. It is the same mistake in both cases.

Just like Sir James Lighthill's report on AGI that started the AI 'winter' of the seventies and eighties,[13] Professor Searle has given us an argument against the next step for strong AGI as well. In both cases, AGI and strong AI, we should note that they are a long way off and currently narrow, or weak, AI is here and working. Neither AGI or strong AI are a requirement for AI. We, as humans, set the goals for our AI to achieve. Working within the EU ethical guidelines for AI points us in the right direction for our AI to have a human-centric ethical purpose, to be trustworthy and technically robust. This approach echoes the work of Professor Max Tegmark, who emphasises the necessity of goals when making explicit what goals machines will achieve.[34] What this means is that any explicit AI machine goal should be focused on achieving goals that are aligned with human goals. He shows that we need machines to align with our goals.

3.2.3 Human plus machine systems

Rather than thinking of either a machine or a human undertaking a task, surely the future is one with humans and machines in harmony? Or, as Stuart Russell emphasises, human compatible AI – this concept is a proposal to solve the AI control problem. Russell's three principles are:[1]

1. The machine's only objective is to maximise the realisation of human preferences.
2. The machine is initially uncertain about what those preferences are.
3. The ultimate source of information about human preferences is human behaviour.

Machines designed in this way will control the development of AI. Ideally, humans will move to higher value work or have more time to enjoy being human. These principles are new, and it's not yet clear how they will pan out; however, Stuart Russell suggests they are futureproof.

Practical examples of human plus machines can be easily imagined. A person using an exoskeleton is able to undertake tasks they weren't previously capable of, perhaps allowing them to walk or lift heavy weights. We have search engines – academics can search literature in seconds, something that would have taken years, if not decades, 40 years ago. In medicine, brain interfaces and the study of the human brain is allowing doctors to better understand stroke recovery. Indeed, we may be able to test for consciousness in patients that we can't communicate with. The idea of humans and machines together gives us super-human capabilities.

3.3 THE CHALLENGES OF ARTIFICIAL INTELLIGENCE

AI is challenging because it is learning from experience and, of itself, learning is a challenge, albeit one that humans rise to continually. Machines can help us with learning from experience.

> Learning from experience does not include automation, if we take the Merriam-Webster definition of automation as:
>
> > Automatically controlled operation of an apparatus, process, or system by mechanical or electronic devices that take the place of human labour.[35]
>
> We see there is no learning from experience here; it's a production line, just like one that produces cans of baked beans in tomato sauce. Beans, sauce and aluminium are fed into the production plant and cans of beans come out the other end. In computing we might write some code to automate a task, such as backing up our work; again, this is not AI or ML. There is no learning from experience.

If we consider AI or ML, we can think of a typical project having the following challenges linked to functionality, software and hardware:

- What **functionality** do we need to learn?
    - collect data;
    - prepare data;
    - algorithms to learn from the data;
    - present data;
    - deploy the AI (product, service, publish).
- What **software** do we need for the above?
    - write our own (e.g. programming languages: R, Python, SciKit-Learn, TensorFlow, C++, Obj C, Fortran, object orientated, etc.);
    - open source;
    - commercial.
- What **hardware** do we need for the software?
    - tablet or smartphone;
    - desktop computer;
    - computer cluster;
    - high-performance computer – cloud service;
    - actuators and sensors – interact with the environment.

These different challenges highlight the nature of AI. It draws on many subject areas and is a universal subject. Practically, we will face mathematical and engineering challenges. We must do this ethically and follow existing laws, for example data protection. Fortunately, we are most likely doing this already and, if we identify where the learning from experience is, we are some way to defining what our challenge actually is. Just defining the challenge doesn't mean it can be solved, but it's a start.

The subjective and human part of any project is always a challenge. We will need to find the right skills, work with and/or around organisational attitudes and possibly even overcome a fear of AI. As Robert Dilts' logical levels of change show us (see Figure 1.2), we will need to build rapport with others and respect and understand their values to make our AI project a success. AI and ML are based on the scientific method and it's more than likely the soft human subjective skills that will cause the most challenges.

3.3.1 General ethical challenges AI raises

As part of the EU's guidelines on ethical AI, we must understand the ethical challenges AI raises. This is by no means a simple subject; we must consider, for example:

- How do we deal with the ethics?
- Who is responsible?
- What does the law say?
- What about human rights?
- What about machine rights?

With AGI and strong AI, robotics becomes another ethical question. General approaches to robotic AI have been published by the Engineering and Physical Sciences Research Council.[36] Other organisations are preparing AI principles also; these include the International Organization for Standardization.[37] Key themes coming out from various organisations include transparency, accountability and care taken when considering bias, diversity, inequalities and unemployment. The themes link back to good ethics. The EU has built the guidelines for AI around an ethical purpose that we can measure using, for example, the United Nations sustainability goals.

3.4 UNDERSTANDING OF THE RISKS AI BRINGS TO PROJECTS

If we assume that we can define a goal for our AI or ML project that has an ethical purpose, then we have to deal with the practical details of what risks AI brings to a project. This means making the goal explicit and then defining a set of tasks, again explicitly. AI projects are engineered, and we are going to have to implement AI systematically. If we are training an AI system or developing an AI algorithm, then we must deal with the mathematics and engineering of our learning from experience. This will possibly be non-linear, complex, involve statistical quantities and so on. It will be difficult to gauge the time it will take to solve this task. These types of problems are risks to your project, so you will need suitably qualified people with experience to help you.

Projects that are non-linear need an iterative approach to solve the underlying mathematical problem. This becomes more involved when the non-linear problem is statistical in nature. An example is predicting the weather. Here, forecasters run 50 simulations (sometimes a lot more) and then assess the statistics of the simulations to understand the likelihood of a particular forecast occurring. We will need to be flexible, adopting an iterative approach to solving the problem.

At every stage of an AI project we are always learning. This can be uncomfortable for some team members who are used to well-organised lists of tasks to do. More complex projects will need a team working together, more than likely in an Agile setting, where regular planning and replannning can take place with the help of a domain expert – remember, they define what is fit for purpose. Communication with stakeholders will be essential, so delivery deadlines are not imposed without a sensible understanding of the iterative approach to projects. If the AI team and stakeholders are not aware of the iterative approach, then this is a major risk to the success of the project.

3.4.1 A typical AI project team

A typical AI project team can have a range of members. It is not easy to generalise because AI is universal. What we know is that an engineered AI project team might need to be:

- 'T shaped' (have high-level understanding of AI or another project specific area, with an in-depth understanding of the core requirement subject area);
- collaborative;
- communicative;
- skilled;
- focused;
- the 'right mindset'.

Typical team members might include:

- product or service owner (business representative);
- data scientist;
- domain expert (ensuring the project is fit for purpose);
- engineer/scientist/mathematician;
- subject matter experts;
- ethicist;
- developers;
- AI trainers;
- AI testers;
- others.

> **WHAT IS A DOMAIN EXPERT OR SUBJECT MATTER EXPERT?**
>
> A subject matter expert 'is a person who is an authority in a particular area or topic'.[38] In the field of AI Expert System software, the term domain expert has been associated with the person who is a specialist in a particular area that forms an expert system; these systems were popular in the 1970s and 1980s. However, with AI now intertwined with ML, the terms are currently used interchangeably and the definitions are blurred.
>
> AI is a truly universal subject, and a AI project team will be made up of a range of subject matter experts or domain experts. Examples of AI subject matter experts or domain experts can include AI and ML experts as well as the experts defining what the AI system is to do. These experts could be business analysts, medical professionals, engineers, scientists, ergonomics specialists, ethicists, historians, artists and so on.

3.4.2 What is 'fit for purpose'?

Fit for purpose means the service or product is well suited to the task, goal or operation it is doing; it isn't too complicated, too costly or takes too long.

A project's output and products must be fit for purpose. EU law states that goods must be fit for purpose. The domain expert defines what is fit for purpose. Practically, the concept of fit for purpose can help us to simplify the learning from experience.

For example, if we are designing a robot that learns how to build a wall in space, the domain expert can save a lot of complication if the spatial accuracy of the robot is millimetres rather than micrometres. In science, the domain expert might specify the accuracy for a calculation to two decimal places, which could mean the ML or AI doesn't need large computational resources.

When working in an Agile team, the domain expert will play a significant role in defining the problem and planning. The domain expert may not have an easy time defining what is fit for purpose in areas that are subjective – satisfying stakeholder expectations is easier said than done.

3.5 OPPORTUNITIES FOR AI

There are numerous opportunities for AI. ML has led the way here and has enabled the next step in AI capability. ML, remember, has been used successfully in many areas for about 80 years. The enablers of the internet, open-source software, cloud computing, the fourth industrial revolution, the IoT are all bringing the next epoch of exploration. The fifth industrial revolution takes us to extreme environments, and the exploration of space and what this has to offer. On a more practical footing, AI and ML offer opportunities in our day-to-day work. Ray Kurzweil, author of *The Singularity is Near*, is an optimistic who looks into and describes the future – a utopian future where we will:[10]

- have the ability to become more human;
- have beneficial use of technologies;
- have a universally sustainable lifestyle worldwide;
- have long-term personal and societal health and wealth gains.

Future AI will give us all super-human insight as well as, perhaps, the means to deliver the utopian ideal. In particular, the singularity is the exponential growth on capability that AI will contribute to.

> In the 1970s, Ray Kurzweil further developed optical character recognition (OCR) technology into what many describe as the first commercial AI product.[39] He created a reading machine that allowed someone with limited vision to have printed text read to them aloud by a machine. Today, we take this for granted and, because we understand this algorithm, we no longer think it's AI. Character recognition is a typical first – 'hello world' – type ML problem we try. Chapter 7 gives more examples of AI in industry and the wider context.

3.6 FUNDING AI PROJECTS – THE NASA TECHNOLOGY READINESS LEVELS

The hype of AI is not enough to write a business case or determine how you will obtain funding for a project. NASA's Technology Readiness Levels (TRLs) provide a measure of the maturity of technology. It was developed to allow NASA to obtain technology cost-effectively from their supply chain. The concept has been adopted by the EU and others, and more details of TRLs can be found at the EU's EARTO project. The TRLs can be described as follows:[40]

- TRL 1 – basic principles observed;
- TRL 2 – technology concept formulated;
- TRL 3 – experimental proof of concept;
- TRL 4 – technology validated in lab;
- TRL 5 – technology validated in relevant environment (industrially relevant environment in the case of key enabling technologies);
- TRL 6 – technology demonstrated in relevant environment (industrially relevant environment in the case of key enabling technologies);
- TRL 7 – system prototype demonstration in operational environment;
- TRL 8 – system complete and qualified;
- TRL 9 – actual system proven in operational environment (competitive manufacturing in the case of key enabling technologies, or in space).

Understanding how mature our AI project is will help us to write a business case and find funders for the project.

3.7 SUMMARY

In this chapter we looked at the benefits and challenges of AI. In doing so, we emphasised what is fit for purpose, a key concept in making sure we have a realistic and credible project. The end of the chapter asked how mature the technology is that we are thinking of developing or adopting. With this knowledge, we are better placed to ask who is going to fund our AI project.

4 STARTING AI: HOW TO BUILD A MACHINE LEARNING TOOLBOX

Our objective in this chapter is to build an understanding of the theory that underpins ML and AI. In doing so, it will give us an appreciation of the education, experience and skill someone developing AI and ML systems needs to have developed.

This chapter is in no way a replacement for education at higher-level institutions and industry. If you are part of an AI or ML team, it is really important to be able to find those people who can work with the more complicated and theoretical aspects of AI. AI is a broad subject and not everyone needs to be applied mathematicians, but it is important have an appreciation that these projects can be complicated and time-consuming. So, our introduction here is designed to inspire but also to remind us that sufficient time should be given to allow members of an AI team to undertake the complex learning required for a project.

4.1 HOW WE LEARN FROM DATA

ML is widely available nowadays and it is an enabler for more involved AI as we accelerate our applications and build intelligent entities. ML is built by defining what functionality we need, writing software to implement that functionality and then constructing the machines (typically digital machines) to obtain our functionality. So, our intuitive and simple model is Functionality, Software and Hardware. AI is about learning from experience and, if we have lots of data, what type of functionality is needed to learn from those data.

4.1.1 Introductory theory of machine learning

At the core of AI is machine learning. In fact, ML is an enabler of AI. Machine learning is most often associated with the popular computation we have today, that is, digital computation. AI is about **learning from experience** and ML is about **learning from data**. We will use this as the basis of our intuitive understanding of what ML is. The most popular machines used today are digital personal computers, computer clusters and large high-performance computers. As we look forward, we will move towards optical, quantum, biological and other types of learning machines to help us.

To start, we need to think about what we are actually doing – we are going to instruct and teach a machine to help us with learning. If we use a digital machine, this means we need to understand our problem in such a way that we can let a machine, say a laptop or smartphone, do the learning for us. The hardware needs software and the software needs an algorithm and data. The algorithm will use mathematics, and we finally have our starting point.

In 2019, Gilbert Strang, a professor at the Massachusetts Institute of Technology, wrote a book on linear algebra and learning from data;[41] he tells us that ML has three pillars: linear algebra, probability and statistics, and optimisation. Much earlier, in the 1990s, Graham, Knuth and Patashnik produced a textbook to help students understand the mathematics needed for computational science; we might think of it as the book that describes algorithms mathematically, how they are defined and how they are understood.[42] These two publications are detailed, rigorous mathematical texts. Tom Mitchell's book on machine learning, as mentioned in Chapter 1, gives a taste of the theory of ML with examples, and is again a detailed, rigorous mathematical text.[23] These books are comprehensive and not for the novice. They are relevant to mathematicians, engineers and physicists who are well versed in these types of disciplines.

AI is more than learning from data; when we think about AI products, we must also think about the world that they operate in, how actuators are controlled and how policies are decided given sensor readings and a percept. Stuart Russell and Peter Norvig's book on AI goes into the theory and mathematics of how we frame our problem beyond just learning from data; they cover ML too. They are also very clear that AI is a universal subject, so can be applied to any intellectual task – learning from experience.[9]

What is common to these technical books on the theory of AI are some key subject areas that are worth gaining an appreciation of. These are: linear algebra, vector calculus, and probability and statistics. Linear algebra is about linear systems of equations; vector calculus is about differentiation and integration of vector spaces – or, more practically, how mathematicians, engineers and physicists describe our physical world in terms of differences and summations and probability and statistics is about the mathematics of understanding randomness.

4.1.1.1 Linear algebra
This section is designed to give an overview of linear algebra; as Gilbert Strang explains, linear algebra is the key subject to understanding data and is especially important to engineering.[43]

We learn from data using linear algebra. In school, most of us learn about systems of equations or simultaneous equations. It is linear algebra that we use to solve equations using computers and it is at the heart of AI, ML, engineering simulation and just about every other subject that relies on digital computers.

A simple example will introduce what we mean. If we want a computer to learn, we have to represent that learning in the form of mathematical operations that a computer can undertake. It is the terrain or geometry that computers can work with. Computers can undertake these mathematical operations, such as add, subtract, divide and multiply, very quickly and accurately. Far quicker and more accurately than us humans. As we mentioned earlier, most of the problems in the real world require an adaptive or iterative approach – and computers are good at this too. In fact, they are so good at this, and quick, we might think it is easy. In doing so, we almost trivialise the amount of work that goes into preparing our machines for learning.

We will use the following set of simultaneous equations to introduce terms that we'll need to be familiar with in linear algebra. Our learning from experience involves solving the following problem:

$$x + y + z = 8,$$

$$10x - 4y + 3z = 13$$

and

$$3x - 6y - z = -20,$$

can we find what x, y and z are?

This can be done by hand using a technique from the 1800s: Gaussian elimination. (Chinese mathematicians were aware of this technique well before Gauss[45].) We could do this by hand with a calculator, a set of log tables or an analogue computer – the slide rule; however, digital computers can do this quicker and more accurately than we can. If our problem has more unknowns, say five or even 1,000, then this would be impossible to do by hand.

To solve this problem with a computer we need to write it in a slightly different way, so we can use linear algebra. We need to define a scalar, a vector and a matrix. This will give us the basic building blocks for our solution. A scalar is a single number that can represent a quantity. Examples are temperature, speed, time and distance. Scalars can be used to represent fields that make up a vector space. A scalar therefore is an element of a vector. An example here is a position given in terms of scalar values of direction x, y and z.

In computing, a scalar is an atomic quantity that can only hold one value. The diagram in Figure 4.1 shows a position vector, **p**, made up of three scalars, x, y and z. Scalars are represented as non-bold characters, for example, x, and vectors are usually bold, for example, **p**, or are bold and underlined, for example, **p**.

The example of a position vector leads to the definition of a vector. A vector has a magnitude and direction. The position vector tells us in what direction it is pointing and how far (magnitude) we have to travel in that direction. The position vector also tells us how far we need to travel in one of the directions, x, y or z. Just as a scalar is an element of a vector, a vector is an element of a vector space.

Vectors can be used to represent physical quantities such as velocity. This is very helpful in science. In AI, we can use these powerful concepts to describe learning mathematically. A simple analogy is that we map out a learning terrain and navigate it. Phrases such as steepest decent or ascent, constraints, hill-climbing or searching now have a physical analogy.

Another linear algebra term that is useful is the matrix. A matrix is an array of elements that can be numbers,

$$\begin{bmatrix} 1 & 3 & 5 \\ 2 & 4 & 6 \end{bmatrix},$$

Figure 4.1 A position vector, p, made of up three scalars, x, y and z

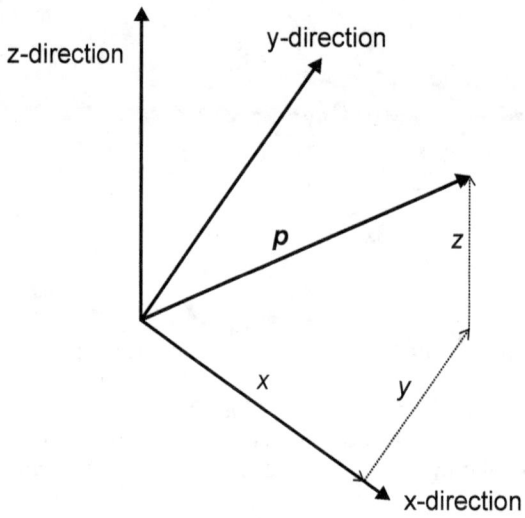

symbols,

$$\begin{bmatrix} x \\ y \\ z \end{bmatrix}$$

or expressions,

$$\begin{bmatrix} e^{-ix} \\ e^{-iy} \\ e^{-ix} \end{bmatrix}.$$

Matrices are versatile and are described by the rows and columns they have. Here, we think of matrices as two-dimensional, but these ideas can be extended to more dimensions. These are called tensors and we need these concepts for more advanced mathematics, engineering and AI. TensorFlow, the open-source software AI library, implies that our learning problems are multidimensional and involve the flow of data.

A matrix is a special form of a tensor and is made up of elements, just like scalars and vectors. The notation we use is described below. A matrix has m elements in its rows and n elements in its columns. You will come across phrases such as an m by n or $m \times n$ matrix. Each element is represented by an index; the first index is a reference to the element in the row and the second reference index is to the element in the column. It is easier to see this pictorially; a 3 by 3 ($m = 3$ and $n = 3$) matrix has elements as follows:

$$A = \begin{bmatrix} A_{1,1} & A_{1,2} & A_{1,3} \\ A_{2,1} & A_{2,2} & A_{2,3} \\ A_{3,1} & A_{3,2} & A_{3,3} \end{bmatrix},$$

with the elements defined by indices as $A_{row,column}$. In this case, the first row is the top row and has the index 1; the second, or middle, row has index 2; and the third, or bottom, row has index 3. In mathematical texts we see matrices written as $A_{i,j}$ where the index that refers to the row is i and the index that refers to the column is j. Some special matrices often used are column and row vectors. This is a 3 by 1 column vector:

$$\begin{bmatrix} 1 \\ 2 \\ 3 \end{bmatrix}.$$

This is a 1 by 3 row vector:

$$[1 \quad 2 \quad 3].$$

So, to recap, we have scalars that are elements of a vector and vectors are elements of vector spaces. Matrices are arrays or numbers, symbols or expressions. If we go to more than two-dimensional arrays, then we call these tensors. We must be careful with tensors and matrices because using them becomes complicated.

Coming back to our system of linear equations solution using a computer, we can rewrite the problem as a matrix multiplied by a vector that equals a vector, as follows:

$$A\mathbf{x} = \mathbf{y},$$

which can be written out fully as the matrix,

$$A = \begin{bmatrix} A_{1,1} & A_{1,2} & A_{1,3} \\ A_{2,1} & A_{2,2} & A_{2,3} \\ A_{3,1} & A_{3,2} & A_{3,3} \end{bmatrix} = \begin{bmatrix} 1 & 1 & 1 \\ 10 & -4 & 3 \\ 3 & -6 & -1 \end{bmatrix},$$

the column vector, $\mathbf{x}$,

$$\mathbf{x} = \begin{bmatrix} x \\ y \\ z \end{bmatrix},$$

and the column vector, $\mathbf{y}$,

$$\mathbf{y} = \begin{bmatrix} 8 \\ 13 \\ -20 \end{bmatrix}.$$

Our solution is to find the values of $\mathbf{x}$; to do this we need to find the inverse of the matrix A, which is

$$\mathbf{x} = A^{-1}\mathbf{y}.$$

Computers can do this effectively and we will discuss some of the commercial and open-source software that we can use later. The answer to this problem is:

$$\begin{bmatrix} 1 \\ 3 \\ 5 \end{bmatrix},$$

or $x = 1$, $y = 3$ and $z = 5$. To check this, put the values of x, y and z into each of the equations and see if you get the same answers.

Scientists, engineers and mathematicians use these techniques to set up problems with large matrices, and have to invert big matrices. This requires high-performance parallel computers, but it can be done. AI draws on these techniques also, and is another user of large supercomputers.

> Strang's *Linear Algebra and Learning From Data* goes into the theory in depth and refers to many excellent examples and references for scientists, engineers and mathematicians.[41]
>
> Graham, Knuth and Patashnik's *Concrete Mathematics* is a book about the mathematics for computer science.[42] It is more detailed in its explanations of algorithms and theory, and introduces statistics and probability. When coupled with Knuth's *The Art of Computer Programming*,[44] they provide an extensive reference literature on the nature of learning using computers, digital and other resources.

If we are to formulate a learning problem that we want a digital computer to solve, then we need to understand linear algebra. There are already multi-billion-dollar industries (finite element analysis, computational fluid dynamics, business analytics, etc.) using these techniques.

Armed with machines that we can use to enhance our learning, we now move onto the next core topic, that of vector calculus.

4.1.1.2 Vector calculus

Vector calculus allows us to understand vector fields using differences and summation. Or, to use the mathematical terms, differentiation and integration. A good example of a vector field is the transfer of heat. Heat can move in three directions, and we observe that it moves from hotter sources to cooler sources. So, in a simple model, we can think that the amount of heat that moves between two bodies depends on the difference in temperature. If we monitor this over a long period of time, we can add these measurements up and calculate the total amount of heat that has been transferred. This is vector calculus; it allows us to represent problems that we encounter day-to-day, from the Earth's electromagnetic field, the movement of the oceans to the motion of vehicles on a road. When we think of a learning agent, we need to represent the state of the world. When we use the analogy of a learning problem as a terrain, then vector calculus will help us to understand how our AI is learning. In particular, how quickly it learns.

A vector is made up of scalar elements and a vector is part of a vector space. The symbol for differentiation is

$$\frac{d}{dx},$$

and integration is

$$\int dx.$$

Mathematicians, scientists and engineers all work with partial and ordinary differential equations to solve their problems. Vector calculus is the cornerstone building block. If they are really lucky, there is an analytical solution to the problem; more often than not, this is not the case and they revert to numerical solutions. We are guided back towards using linear algebra.

Dorothy Vaughan used these techniques in the 1950s to help NASA land a vehicle on the moon and return it safely.[46] More generally, we don't need to think about vector calculus applying only to space, that is, Euclidean space of x, y and z directions. The ideas and concepts can be generalised to any set of variables, and this is called multi-variable calculus. We can effectively map out a space made up of different variables and use differentiation and integration to understand what is happening.

If we were training a NN, we could map out the training time in terms of number of layers, number of nodes, bias, breadth and width of the layers, then understand how effective our training is using vector calculus. For example, we might ask what gives us the minimum error in the quickest time.

A quick example of vector calculus is useful here. Let's imagine we want to know how much heat is being lost through the wall of our house; this is shown schematically in Figure 4.2.

Figure 4.2 Example problem for vector calculus

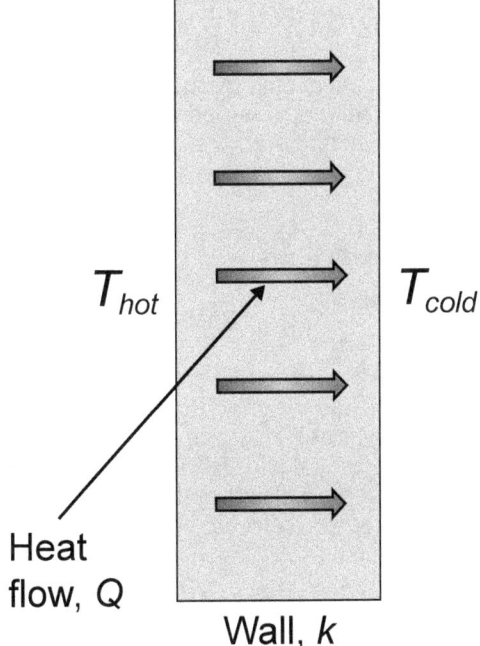

In the room in the house the temperature is hot, this is T_{hot}, and outside the building is the outdoor temperature, and this is cold, T_{cold}. We know that heat flows from hot sources to cold sources. We can also see that it is conducted through the wall. The amount of heat conducted through the wall is proportional to the thermal conductivity of the wall. The conductivity of the wall is represented as k. We would like to know how much heat we are losing through the wall; we'll give this quantity the variable Q.

In words, the amount of heat we are losing through the wall, Q, is calculated by multiplying the thermal conductivity of the wall, k, by the temperature gradient through the wall. We also need to incorporate the surface area of the wall, A. The bigger the wall, the more heat we lose. What is the gradient? This is the temperature difference divided by the wall thickness, which is given the variable δ. Practically, this is what we are doing:

$$Q = -Ak \frac{(T_{hot} - T_{cold})}{\delta}.$$

We insert the negative sign to show that heat flows from hot to cold sources and ensure we obey the second principle of thermodynamics. In vector calculus we write this as:

$$Q = -Ak \frac{\partial T}{\partial x}.$$

The notation is more elegant, and the derivative is written in a form called the partial differential, $\frac{\partial}{\partial x}$. This is saying that we are only interested in the gradient in the x-direction. We mentioned earlier that heat can flow in three spatial directions and, if we want a more complete picture of the heat transferred by conduction, we can rewrite our equation as:

$$\mathbf{Q} = -kA \left(\frac{\partial T}{\partial x} \mathbf{e}_x + \frac{\partial T}{\partial y} \mathbf{e}_y + \frac{\partial T}{\partial z} \mathbf{e}_z \right),$$

which is telling us that the heat conducted through the wall is the thermal conductivity of the wall multiplied by the gradients of temperature in each of the spatial directions. The terms $\mathbf{e}_x$, $\mathbf{e}_y$ and $\mathbf{e}_z$ are unit vectors that have a magnitude of 1 and point the x, y and z directions respectively.

Vector calculus is very powerful and elegant; it also lets us express our equation in a simpler form:

$$\mathbf{Q} = -Ak\nabla T.$$

What we notice here is that the heat is now a vector, $\mathbf{Q}$, because we are looking at three dimensions. k and T are scalars and the operator ∇ is a vector. Our objective here is to gain an appreciation of the power and elegance of the fundamental mathematics that underpins AI. We can extend vector calculus to vector fields, not just scalar fields in our example. Applied mathematicians, physicists and engineers use these techniques daily in their jobs. When we couple vector calculus with linear algebra, we can use digital computers and machines to help us solve these equations.

Models of the state of the world (agent's internal worlds) or ideas such as digital twinning use these techniques. These ideas work quite well on engineered systems where we have some certainty on the system we are working with. The real world, however, is not perfect and in AI we also need to deal with uncertainty or randomness. This is the next mathematical topic that AI needs – probability and statistics.

4.1.1.3 Probability and statistics

Probability and statistics are vital because the world we live in is random. The weather, traffic, stock markets, earthquakes and chance interfere with every well thought out plan. In the previous sections we introduced linear algebra and vector calculus, which are elegant subjects but also complex and need the careful attention of a domain expert. Randomness adds another dimension to the types of problem we might face with AI.

Applied mathematicians, engineers and physicists all use statistics to extend the capabilities of linear algebra and vector calculus to understand the world we live in. This is easier to observe in the diagram at Figure 4.3. Some examples are helpful here. The low-randomness, low-complexity learning is what we can work out as humans, perhaps with a calculator. For example, timetables – how long it takes for a train to travel from London to Kent. We can make these calculations easily. The low-randomness, high-complexity quadrant is learning that involves complex ideas or theory, like linear algebra or vector calculus. An example of this is how Dorothy Vaughan used computers and Fortran to calculate the re-entry of a spacecraft into Earth's atmosphere. Low-complexity and high-randomness learning is where we can use simple ideas to understand randomness. Examples of this are games of chance, such as card games, flipping a coin or rolling dice. Physical examples are the shedding of vortices in a river as it passes around the leg of a bridge. High-complexity and high-randomness is learning that involves complex ideas that also include high randomness; examples are the weather, illness in patients, quantum mechanics, predicting the stock market. They often involve non-linearity, discontinuity and combinatorial explosions.

Dorothy Vaughan's work using computers to calculate the re-entry point of a spacecraft into the Earth's atmosphere was an elegant example of using a computer to iterate to find a solution – something that we use a lot in learning. Learning is non-linear and iterative. Dorothy took a complex problem and simplified it so a computer could guess repeatedly and find a solution. We can think of this as a heuristic that allowed a digital machine, using Fortran, to guess a solution. This is shown in Figure 4.4.

Figure 4.3 Complexity and randomness

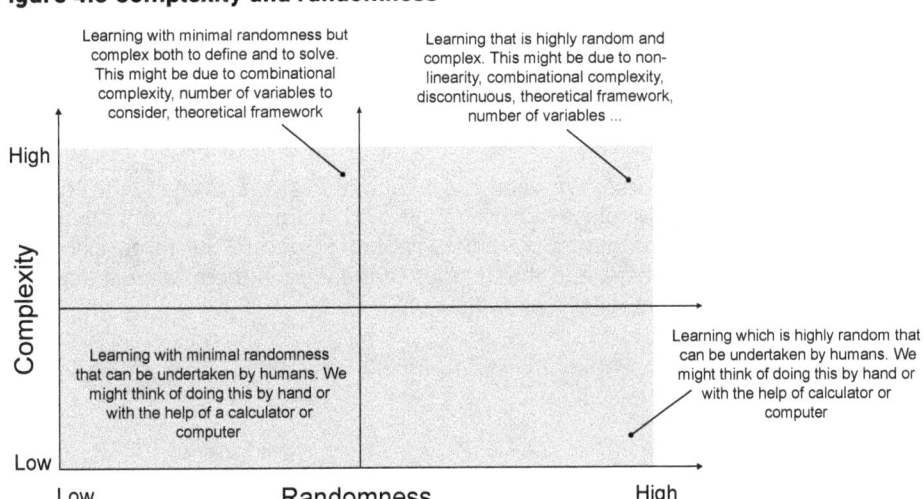

Figure 4.4 Heuristics can simplify complexity and randomness

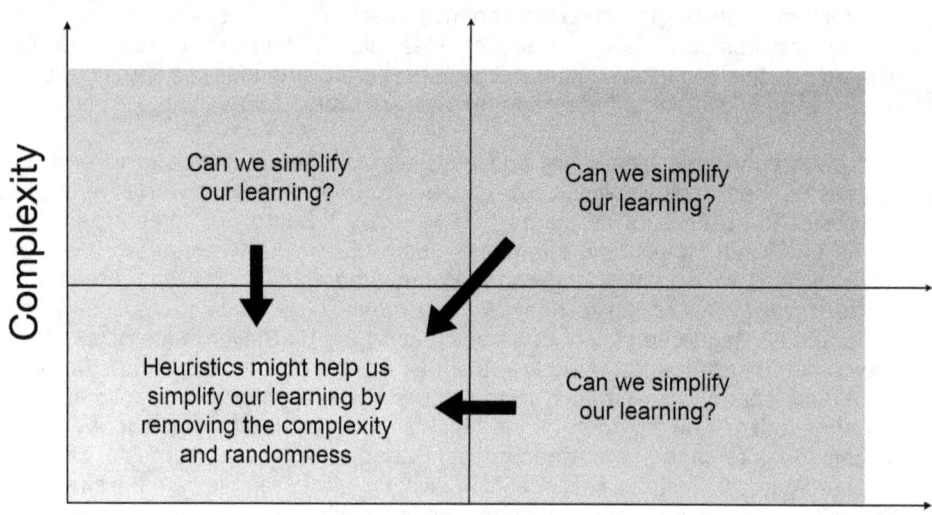

Statistical analysis of large volumes of data is a modern-day success of digital computers. Without them we would not be able to analyse the information encoded in our genetics. Business analysts are making our activities more effective and efficient, leading to better supply chains that are integrated. An intuitive understanding of statistics and probability will help those working in AI. In AI, of particular importance, is inference and the ability to infer something about data. This started some time ago with Reverend Thomas Bayes. He came up with a theory (Bayes' theorem) that allowed us to infer the probability or likelihood of an event happening given knowledge of factors that contribute to the event.[47] We need to remind ourselves that the factors may or may not cause the event, but we have noticed that, if they occur, it could also mean that the event will happen. For example, it rains when there are clouds in the sky (50 per cent of the time), the temperature is 19°C (33 per cent of the time), the pressure is ambient (22 per cent of the time) and so on. We can now ask ourselves what would be the probability of it raining if there are no clouds in the sky and the temperature is 19°C. Can we infer this from the data we have already?

To understand Bayes' theorem we need to go back to basics and start with Russian mathematician, Andrey Nikolayevich Kolmogorov.[48] Kolmogorov's contribution to probability and applied mathematics is astonishing. He wrote the three axioms of statistics. To understand them, and the theories that underpin them, we will use Venn diagrams, a pictorial way to represent statistics.[49]

Say we are interested in a sample space, S, of events, A_i. The space is defined as the union of all the events:

$$S \equiv U_{i=1}^{n} A_i.$$

which is a set of all possible outcomes. We need to be able to define what happens if an event never happens (is impossible), or if an event always happens (is certain) or, finally, is random (sometimes occurs). From this space of events, we would like to able to understand how these events are related to and depend on each other. To do this, Kolmogorov states that in order to define the probability of an event, Ai, occurring we need three axioms.

The first axiom is:

$$0 \leq P(A_i) \leq 1,$$

which defines that the probability of an event, A_i, happening lies between zero and one. If it is impossible, then it is false and has zero probability. If it is certain to occur, then it is true and has a probability of one.

The second axiom is:

$$P(S) = 1,$$

which means that the union of the sets is always equal to 1. So, if $P(A)$ is true then $P(A) = 1$ and if $P(A)$ is false then $P(A) = 0$. Sometimes this is explained as at least one event must have a probability of one.

The third axiom is:

$$P(A_1 \cup A_2) = P(A_1) + P(A_2),$$

where the events A_1 and A_2 are mutually exclusive. This means that if A_1 occurs then A_2 cannot occur. Or, if A_2 occurs, then A_1 cannot occur. The third axiom leads to the addition law of probability, which deals with events that are not mutually exclusive, and this is:

$$P(A) + P(B) - P(A \cap B) = P(A \cup B),$$

where $P(A)$ is the probability of A, $P(B)$ is the probability of B, $P(A \cap B)$ is the probability of A and B and $P(A \cup B)$ is the probability of A or B. We can write this out as the probability of A or B can be found by adding the probability of A, $P(A)$, to the probability of B, $P(B)$, and subtracting the probability of A and B, $P(A \cap B)$. This makes more sense when we look at Venn diagrams.

Let's say we have a bag of letters made up of A, B, AB and BA and this bag of letters defines our sample space, S. We'd like to know what the probability is of certain events occurring. The total number of letters we have is 100, with 43 As, 36 Bs, 15 ABs and 6 BAs.

So, we can build the Venn diagram as follows. Our total space is 100 letters. We build an area of 100 units and add to it the areas of the sets for the probability of pulling an A or a B out of our bag. The probability of pulling an A out of the bag is:

$$P(A) = \frac{43 + 21}{100} = 0.64,$$

and pulling a B out of the bag is:

$$P(B) = \frac{36 + 21}{100} = 0.57.$$

We have letters in our bag made up of ABs and BAs, so pulling an A or a B out of the bag is not mutually exclusive. So, just to cover the axioms, it's worth noting that if our bag contained only Bs and no As, then the Venn diagram would look like the top diagram in Figure 4.5, and if our bag of letters contained only As, then the Venn diagram would look like the bottom diagram in the figure.

Figure 4.5 Venn diagrams for false {none of the space S is covered} and true {all of the space S is covered}

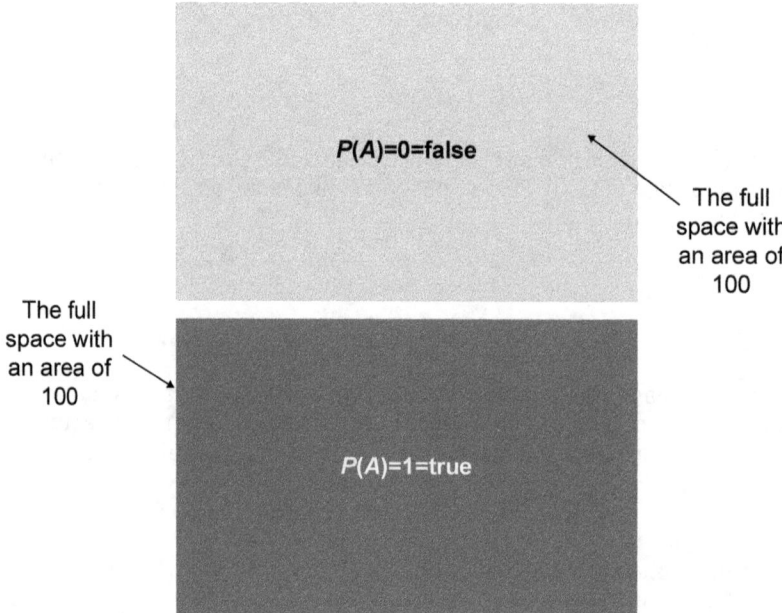

The probability of something being true is 1, and the probability of something being false is 0.

Figure 4.6 shows the Venn diagram of the probability of pulling an A (top diagram) or a B (bottom diagram) from our bag. We see that the $P(A)$ and $P(B)$ regions overlap. With our first look, it seems odd because, when we add up the probability, $P(A) + P(B) \neq 1$. This is because we can pull out a letter of ABs or BAs, so they are not mutually exclusive. Figure 4.7 shows the probability of pulling a letter A or letter B out of our samples. We see that these regions overlap too.

Figure 4.6 Venn diagrams for the probability of finding a letter A, $P(A)$ (top) and the probability of finding a letter B, $P(B)$ (bottom); A and B are not mutually exclusive

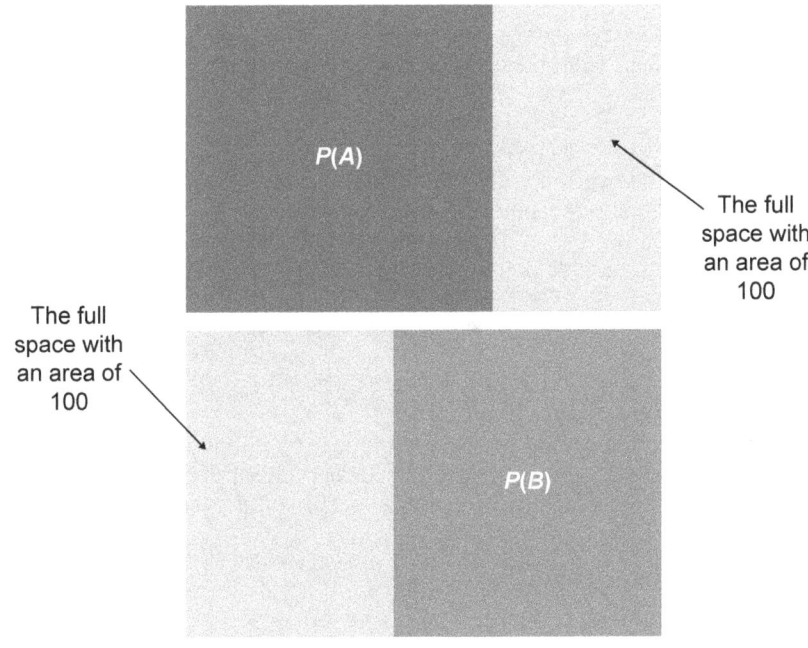

Figure 4.7 Venn diagram where A and B are not mutually exclusive. Centre region is accounted for twice in the addition law of probability

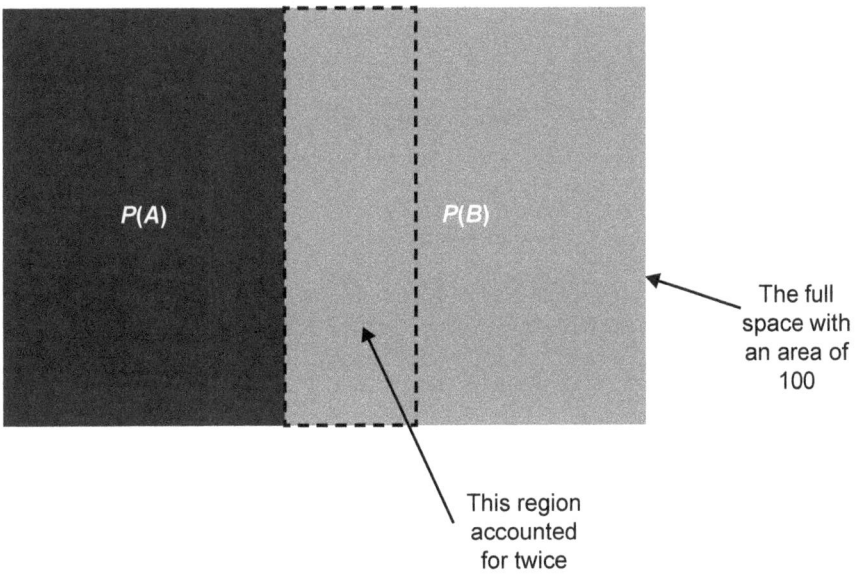

We can see why, from the addition law of probability,

$$P(A) + P(B) - P(A \cap B) = P(A \cup B).$$

In our bag of letters, we have letters that are ABs or BAs, so it is possible that we can obtain an A and a B. In our examples, the probability of obtaining an A and a B are:

$$P(A \cap B) = 0.21,$$

which is the area of the overlapping region. This has been accounted for twice. The addition law of probability can be checked. If we asked the question 'What is the probability of obtaining an A or a B?', our answer should be 1. All of our letters are either an A or a B. Using the addition law of probability, we can check our understanding:

$$P(A) + P(B) - P(A \cap B) = P(A \cup B),$$

$$0.64 + 0.57 - 0.21 = 1.$$

It's also helpful to define conditional dependence, conditional independence and statistical independence. If our events occur occasionally, and are therefore random, we can ask: What is the probability of event, A, occurring if event, B, occurs? This is written as

$$P(A|B).$$

The definition of conditional probability is:

$$P(A|B) \equiv \frac{P(A \cap B)}{P(B)}.$$

We can see what this means via the Venn diagram in Figure 4.8.

The definition of independence is:

$$P(A|B) \equiv P(A),$$

and

$$P(B|A) \equiv P(B),$$

which means that the occurrence of event A does not affect the occurrence of event B. By defining what they are, we can test if $P(A)$ and $P(B)$ are independent.

What this is telling us is that, on a Venn diagram, the area of $P(A)$ is equal to the area of $\frac{P(A \cap B)}{P(B)}$. Just to recap, if:

$$P(A|B) = P(A),$$

Figure 4.8 Venn diagram defining the conditional probability, P(A|B) of event A given event B. The black circle is the set of events A, and the dark grey circle is the set of events B

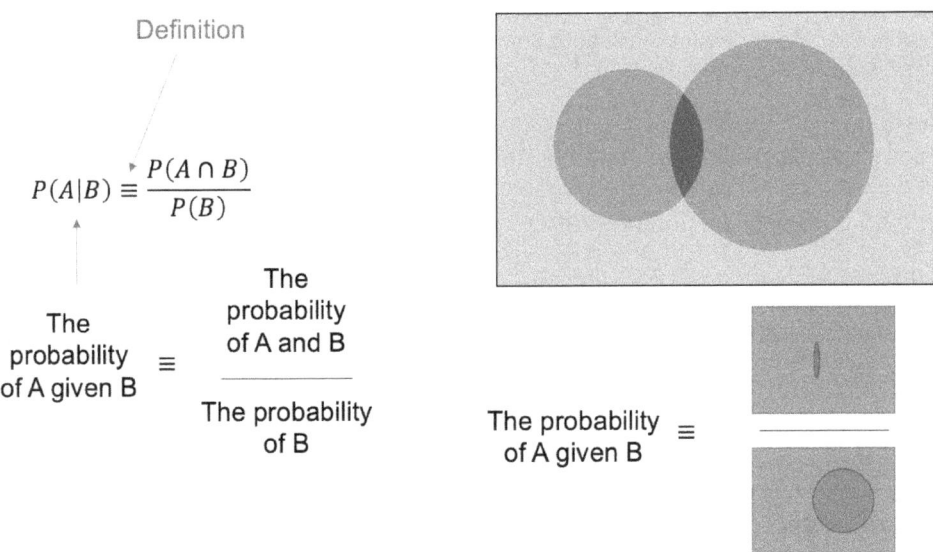

that is, random event, A, does not depend on the random event, B, and is therefore independent, then

$$P(A|B) = \frac{P(A \cap B)}{P(B)} = P(A)$$

or

$$P(A \cap B) = P(A)P(B).$$

Even though this appears intuitive in practice, when we extend this to multiple variables, it is complicated and requires careful thought. Computers are a good way to organise this.

Networks can be built using Bayes' theorem:

$$P(A|B) = \frac{P(B|A)P(A)}{P(B)}.$$

These axioms and definitions can be used to build an understanding of how random events are related to each other and to an outcome or event. This can be very useful in medical diagnosis, decision making and other areas such as the prediction of defective products or maintenance. These ML techniques involve Bayes' theorem and the law of total probability.[50] They build networks or graphs – Bayesian networks[51] – and are graphical representations of how an event is influenced by possibly known variables or causes.

In AI there are two main types of statistical learning, regression and classification.[52] Classification tries to classify data into labels, for example name a person in a picture. Regression tries to find a relationship between variables, for example the relationship between temperature and heat transfer. This brings us nicely to how we represent random data and how we generate random numbers. Perhaps we have a limited set of statistical data and want more data points? This requires an understanding of how we represent random data.

Random data comes in two types, discrete and continuous. An example of discrete random data is the score from the roll of a dice. An example of continuous random data is the output from a pressure sensor recording the pressure on the skin of an aircraft's wing. It is useful to know how these random data are distributed.

For discrete data we use a probability mass function, and this is shown in Figure 4.9. It is sometimes called the discrete density function. In this example we have a histogram showing the possible outcomes of rolling two dice. They are discrete values because we can only obtain the following scores from adding the two:

2, 3, 4, 5, 6, 7, 8, 9, 10, 11 and 12.

Figure 4.9 Example of a probability mass function of rolling two dice (reproduced from https://commons.wikimedia.org/wiki/File:Dice_Distribution_(bar).svg under Wikimedia Commons licence)

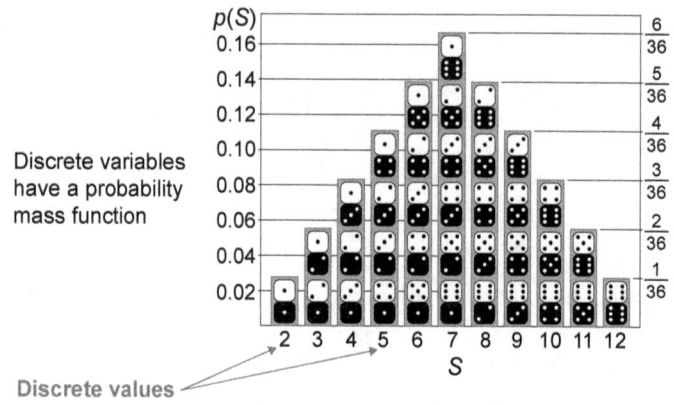

We can see that there is one way to obtain the score 2 and six ways of obtaining the score of seven. The probabilities of obtaining each score, 2 to 12, add up to 1.

Continuous random data are represented by probability density functions (PDFs).[53] Probability is now thought of as being in a range, say in between two values. So, from our example of the pressure acting on the skin of an aircraft's wing, what is the probability of the pressure being between 10Pa and 15Pa above atmosphere? Or we can think of the traditional way we grade exam scores; different ranges are associated with the grades received by the people taking the exam. This is shown in Figure 4.10. Here we

Figure 4.10 Example of a probability density function of continuous data (reproduced from https://commons.wikimedia.org/wiki/File:Standard_deviation_diagram.svg under Wikimedia Commons licence)

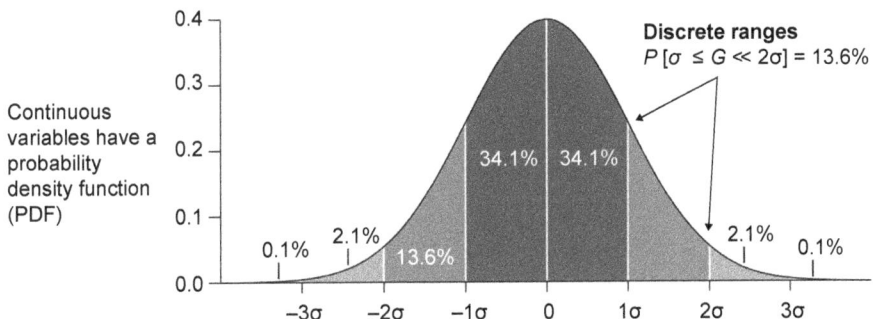

see a standard Gaussian curve and the mean score is in the middle. Particular grades are associated with the bands in the curve that define a score range.

PDFs come in many forms and have different measures to describe them. A uniform and Gaussian (also called normal, Gauss, LaPlace-Gauss) PDF, along with their equations, can be seen in Figure 4.11. By describing our discrete or continuous random data via a distribution, we could also generate random numbers, and we will come back to this later. There is another important idea that can be described by example.

If we have five dice, and use them to generate random scores, we could ask what the distributions would look like; in particular, what happens as we increase the number of dice that contribute each set of scores we generate. This experiment is shown graphically in Figure 4.12. Rolling one die, as we expect, is a uniform distribution because each score is equally probable. When we roll two dice, the distribution is no longer uniform. As we increase the number of dice, the distribution converges onto a normal distribution. The theorem that describes this is the central limit theorem.[54] The central limit theorem suggests under certain (fairly common) conditions, the sum of many random variables will have an approximately normal distribution. So, if an event is a combination of lots of random data with different distributions, it is likely our event's distribution will be Gaussian or normal.

From our intuitive understanding of statistics and probability we can express random data in terms of distributions and how random events relate to each other. However, we are missing how we generate random numbers or random data. Computers are explicit, linear, precise and deterministic. Fortunately, they can produce pseudo-random numbers, and the algorithms to do this are already written.[44, 55] If we have incomplete data, but can somehow determine the probability distribution from the data we have, then we can generate random numbers with the same distribution.

Section 4.1.1 has built up an intuitive understanding of the pillars of ML and AI. Domain experts in these areas are needed if you are working with AI at an algorithmic or

Figure 4.11 Uniform and Gaussian probability density functions of continuous data
(reproduced from: (top) https://en.wikipedia.org/wiki/Uniform_distribution_(continuous)#/media/File:Uniform_Distribution_PDF_SVG.svg; (bottom) https://commons.wikimedia.org/wiki/File:Normal_Distribution_PDF.svg under Wikimedia Commons licence)

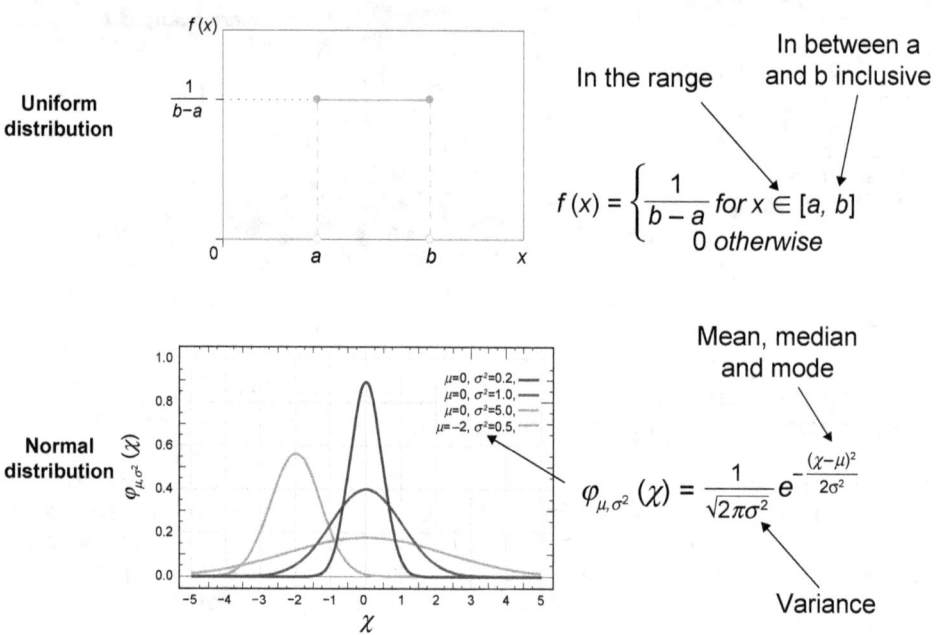

training level, so you will more than likely need to have these people in your team. What we have tried to show in this section is the level of mathematical understanding someone implementing AI needs to have when working with AI software. A programmer just learning Python, for example, is not prepared to develop AI algorithms; however, if a programmer is confident in the material in Section 4.1.1, then some higher-level mathematical training will make their job much easier.

The case study in this section gives an integrated example of how vector calculus, linear algebra and probability and statistics are used at a scientific level. Vector calculus allowed us to understand the fundamental phenomena mathematically; the linear algebra allowed us to build a simulation (we might think of this as a digital twin) and generate data; and probability and statistics allowed us to generate and understand the complex non-linear nature of the data we had generated. In an AI project we might not have such tight control of the data generation, as we might have data from the real world.

4.1.2 Typical tasks in the preparation of data

Preparing data is an important aspect of all work with computers, not just AI or ML. In AI, the phrase 'learning from experience' is paramount. There is no guaranteed, sure fire

Figure 4.12 An example of the central limit theorem using five dice and rolling set of scores. Top left is a set of scores from rolling one die (n = 1); middle right is a set of scores from rolling five dice (n = 5) (reproduced from https://commons.wikimedia.org/wiki/File:Dice_sum_central_limit_theorem.svg under Wikimedia Commons licence)

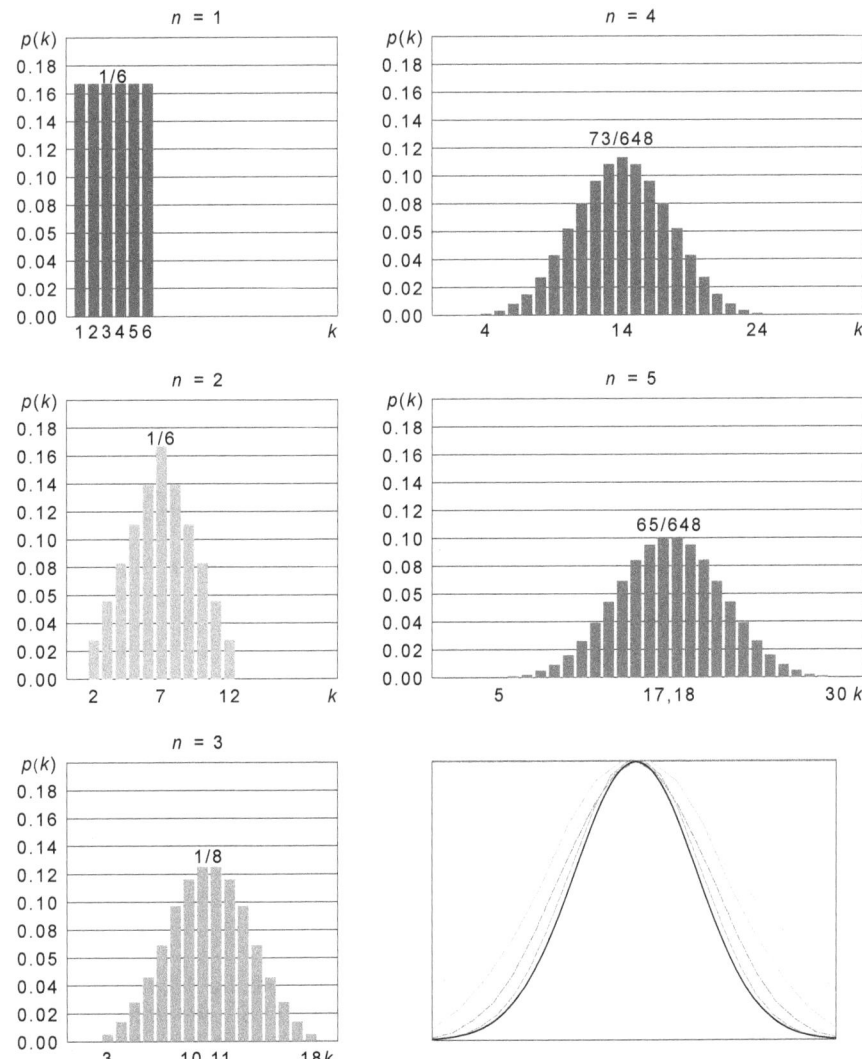

way to ensure your data preparation is right first time. You have to learn from experience and improve your data preparation with each iteration. So, when do we stop? That is something determined by the domain expert – this person sets the standard for what is fit for purpose. An Agile project style lends itself to this very well.

Data preparation is affected by numerous factors and is a science in its own right; it therefore needs experience to undertake. This experience could be learned during the project or you could employ someone already experienced to do this for you.

> The Further Reading list at the end of this book has numerous ML examples that will give you a flavour of the data preparation needed.

We should keep in the back of our minds that if we are working with large data sets, we have the added complexity of obtaining the data, storing them, organising them, keeping the data safe and obeying any laws associated with them – a time-consuming task in itself. Aurélien Géron's eight stages of a machine project is a good place to start.[56] These are:

1. Frame the problem.
2. Get the data.
3. Explore the data.
4. Prepare the data to better expose the underlying data patterns to machine learning algorithms.
5. Explore many different models (or algorithms) and shortlist the best ones for the project.
6. Fine-tune your models and combine them into a great solution.
7. Present your solution.
8. Launch, monitor and maintain your system.

Again, we notice immediately that there is considerable learning from experience involved in these eight stages. They are not sequential, and we may revisit or loop over stages 1–5, 1–7 or 1–8 – going through stages 1 to 8 in one pass will be the exception rather than the rule. We might use smaller data groups so we can learn quicker because we are not processing lots of data. This can be helpful when assessing multiple algorithms as there is more work associated with this. Some learning can be gained by working on smaller subsets at an early stage. It simply reduces the time between iterations.

If we have the luxury of a large data set, we will need to break up the data into training data and test data. This is important for the trade-off, where we want to trade off complexity and error; this is covered in more detail in Section 4.1.4. We might then break up our training data into data bins of different sizes and use those different sized bins to test our algorithms. As we refine our approach, we can then use the test data to see if our learning works on data that have not been seen by the AI or ML system before.

Test data should **not** be used for training your AI or ML system. One of the reasons for doing this is that the algorithm will use the test data while training so we already know how well the algorithm deals with these data. It is essentially a waste of time because we are not adding further information to the system. Independent test data gives us an independent test of the ML we have done.

Once we have our data and they are divided up into appropriate bins, we can then start to clean or scrub the data. This means that we prepare our data in such a way that they give our algorithms the best chance of success.

In AI, we can encounter many types of data, such as numbers, strings, Booleans, or they could be more sophisticated data, such as audio files, movie files, script, text, documents and, perhaps, signals from actuators, sensors – we might even have to generate our own data in simulations. Digital twin is an exciting development that has gained momentum over the last 20 years or so. Data will come to us in many forms, in varying quantities.

Big Data is a term that often describes the vast array of data that can be found on the internet. Big Data has the following characteristics:[57]

- Volume measures the quantity of stored and generated data.
- Variety measures the nature and type of data.
- Velocity measures the speed at which the data is moving.
- Veracity measures the data quality, value or utility.

These are self-explanatory, but the last one recognises that we may have data that are of no use or utility is of low quality. Part of the preparation of data is data cleaning or data scrubbing. As we are going through the stages of a ML project, we must always be aware of the quality of our data preparation. Poor data preparation will give spurious ML results or misrepresent the data we are trying to analyse.

Scrubbing can include:

- reducing the volume of the data we are working on, perhaps removing irrelevant data;
- reformatting data;
- removing incomplete or duplicated data;
- dealing with missing data;
- generating data;
- scaling data so algorithms perform well.

There are many tools and techniques for data preparation and standard libraries to help us, and we need experience to do this well.

As important as data preparation is data visualisation. Visualisation of data (see Section 4.1.5) will also affect how you prepare your data. This is often overlooked until we try to visualise a data set that spans multiple processors on a large supercomputer.

4.1.3 Typical types of machine learning algorithms

If we have prepared our data well and are confident they can be used to test a hypothesis, then we can begin to understand the types of algorithms we employ in ML. What we mean by 'type' is the broad categories that define our algorithms. They are typically organised as follows:[9]

- **supervised learning** – these types of algorithm use labelled (actual solution) data to train the algorithm (see Figure 4.13).
- **unsupervised learning** – these types of algorithm use unlabelled data to train the algorithm (see Figure 4.14).
- **semi-supervised learning** – these types of algorithm use labelled and unlabelled data to train the algorithm (see Figure 4.15).
- **reinforcement learning** – these types of algorithm use the learning of an agent to achieve a goal that is measured in terms of a reward or penalty (see Figure 4.16).

Examples of each of the types of ML are:

- **supervised learning** – classification, regression (linear, logistic), support vector machines, decision forests, neural networks, neighbours;
- **unsupervised learning** – clustering, association rule learning, dimensionality reduction, neural networks;
- **semi-supervised learning** – combinations of supervised and unsupervised algorithms;
- **reinforcement learning** – Monte-Carlo, direct policy search, temporal difference learning.

> Reinforcement learning fits naturally with the AI learning agent schematic. It relates to both products (e.g. autonomous robots) and data analysis. In reinforcement learning, an agent interacts with an environment to achieve a goal. It can also learn how to achieve that goal.

We can also categorise the way an algorithm learns in terms of the data it is working with.

Batch learning describes learning from large data sets. All of the data are used to train and test the algorithm. The computer resources required are governed by the volume, velocity, variety and veracity of data. This learning is done offline. Online learning is undertaken with data in small or mini batches. Learning occurs as data become available – an example is a system that learns from stock market prices.

There will be times when we come across types of learning from examples or learning from examples to build a model; these types of learning are called instance-based learning and model-based learning, respectively. These are possibly the simplest form of learning.

Figure 4.13 Supervised learning

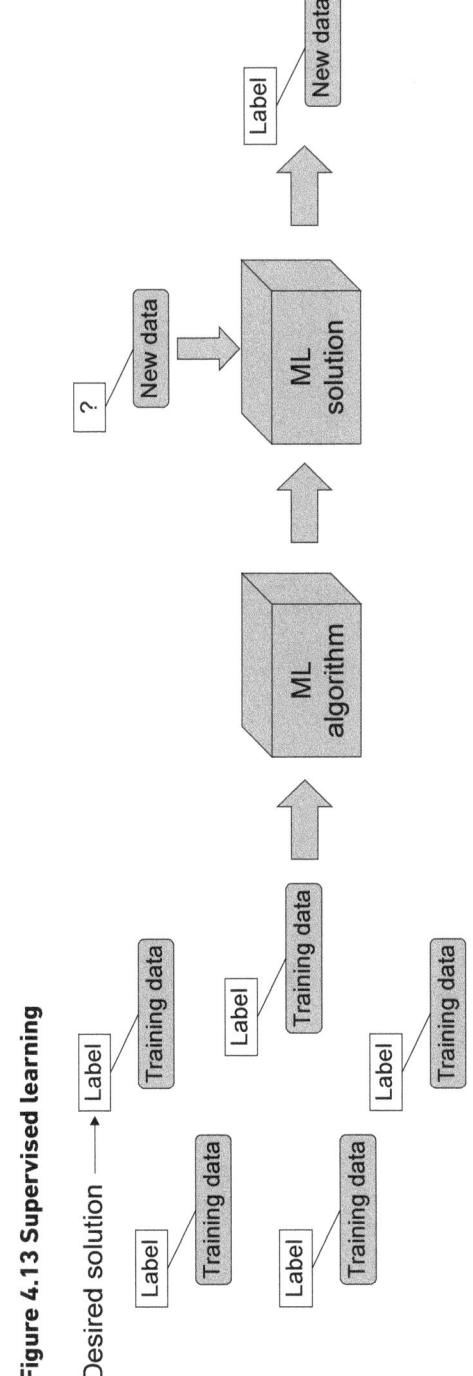

Figure 4.14 Unsupervised learning

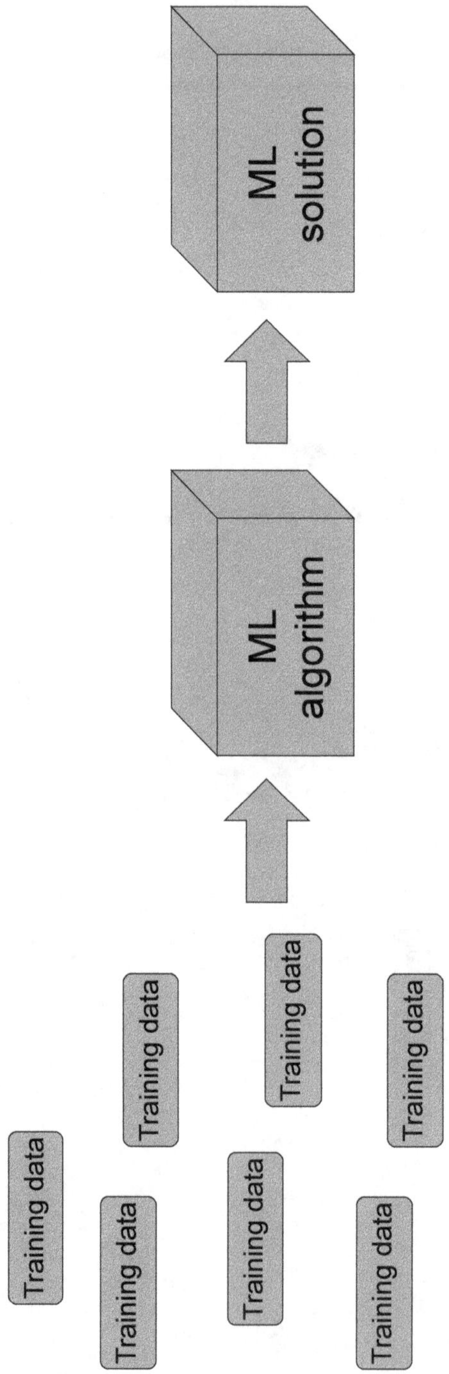

Figure 4.15 Semi-supervised learning

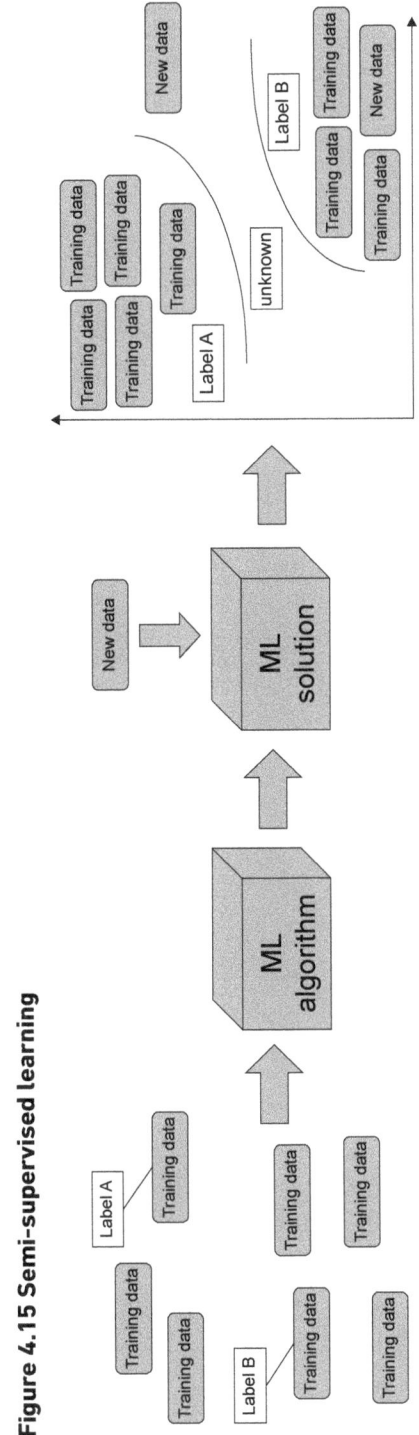

Figure 4.16 Reinforcement learning

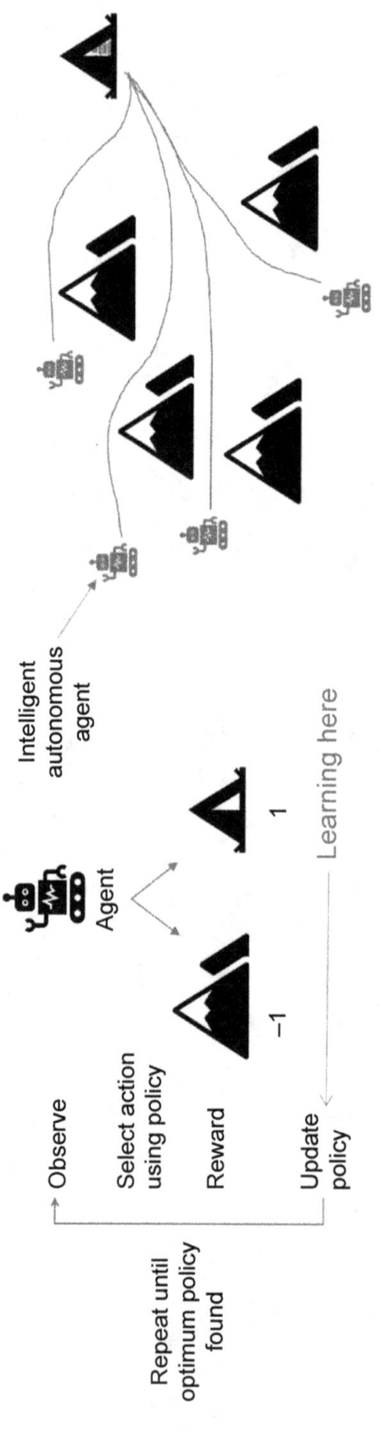

4.1.4 Hyper-parameters, algorithms and the trade-off – under- and over-fitting

Our learning from experience involves assessing the efficacy of an algorithm to test our hypothesis. We will try different types of algorithm and adjust the algorithms' tuning parameters to get them to perform the best they can. The tuning parameters are called hyper-parameters and they are used to set up our algorithms; or, to put it another way, they tell the algorithm how to learn. Examples of hyper-parameters include layers in a decision tree or NN, the order of a curve in regression, the number of nodes in a NN, the type of activation function, or the accuracy of a classification.

We will not know before we start what our hyper-parameters need to be to obtain the most effective algorithm. This is part of our learning from experience as we build our ML system. There are also many ways to assess the efficacy of an algorithm. The most common one used is the trade-off between complexity and accuracy. To do this, we need to understand under- and over-fitting.

These are exactly as their names imply: over-fitting the model over-fits the data too well, and under-fitting under-fits the data and does not capture the true nature of them. You can reproduce over-fitting and under-fitting on a spreadsheet by simply fitting different curves to a data set.

With over-fitting (see Figure 4.17) we have a data set that represents a linear straight line. This could be from experimental measures of temperature through a metal bar. The good fit straight dashed line in the graph shows the best fit to the data set. There is only a small variation of the data when compared to the best fit straight line. If we try to be too clever and use a more complex mathematical fit, say a parabola or a higher order polynomial, then we will begin to see larger errors. In the curve fitting example, the higher the order of the curve, the more oscillatory the data fit we obtain. These show up as large errors away from the actual data points we use to train our ML solution. In fact, if we have nine data points, we need a ninth order polynomial to obtain an exact match of the data points with our polynomial. However, this ninth order polynomial could oscillate widely and be a very poor fit away from the data points. To recap, over-fitting means our model or algorithm is too complicated for the data we are working with.

Figure 4.17 A simple example of over-fitting

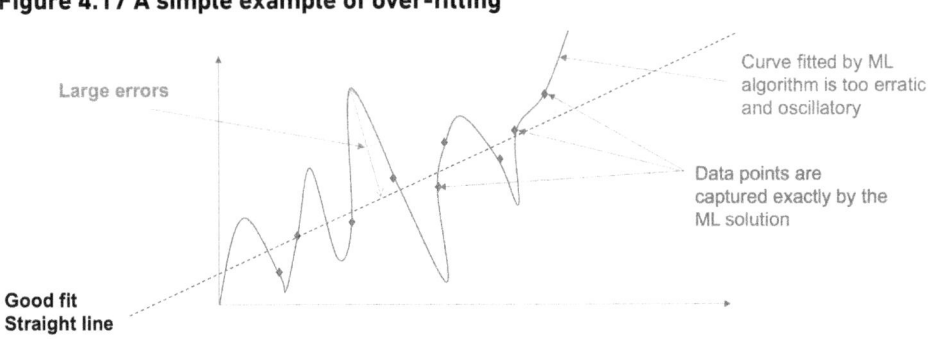

Under-fitting, on the other hand, is the opposite; it tries to fit a curve (using our simple curve idea) that is not complicated enough. In Figure 4.18 we see that our data set is well represented by a parabola (dashed line) or a second order curve. If we try to fit a simple linear straight line to the data, we again see large errors.

Figure 4.18 A simple example of under-fitting

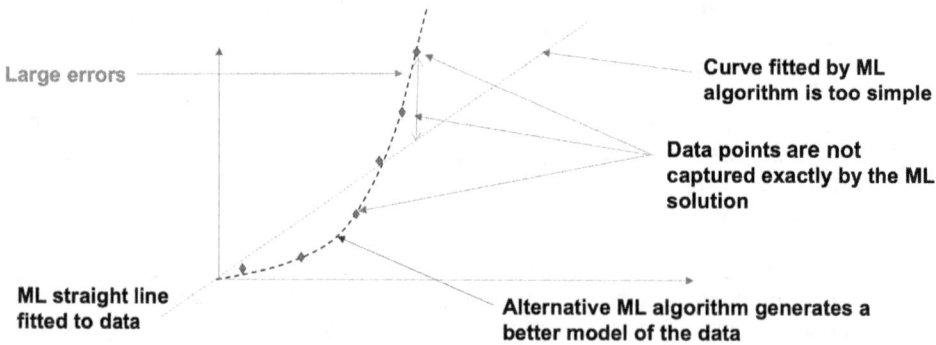

So, our algorithm suffers from two types of error: bias error and variance error. The bias is how far your prediction is from the actual data. Variance is how scattered around our prediction is. With over-fitting, we see a high variance around and low bias with our training data. With under-fitting, we see a low variance around and high bias with our training data. The trade-off, shown graphically in Figure 4.19, minimises the error by finding the best level of complexity. This is not the only way to assess the efficacy of an algorithm. There are many other ways, and there are numerous examples to be found in the books listed in Further Reading at the end of this book. We should also note in passing that the trade-off uses test data that have not been used to train our algorithm.

Remember that the learning from experience within a ML project involves the whole team. The domain expert is the person that defines what is fit for purpose.

4.1.5 Typical methods of visualising data

Visual representation of data is one area where a focus on AI or ML can enhance how you organise and present data. The benefits of good visualisation must be a priority as, ultimately, we are judged by clients, stakeholders and/or users on the deployed AI or presentation of results. Visualisation is a fundamental enabler of learning from experience for everyone on the team. Figure 4.20 again shows the stages of a ML project and highlights how visualisation can help with learning from experience at every stage (see Géron's eight stages in Section 4.1.2).

If we can visualise data from the start of a project, we magnify our learning – a picture paints a thousand words. In addition, we are already organising our data so we can manage them and use them. If we are in a highly regulated profession, then early engagement with stakeholders and regulators might aid the acceptance of work and results.

Figure 4.19 Algorithm assessment, the trade-off and finding the balance between under- and over-fitting

Trade-off
Under-fitting
Low variance
High bias
Over-fitting
High variance
High bias
High
Model error
Low
Low
Model complexity
High
Error between model and test data
Error between model and training data

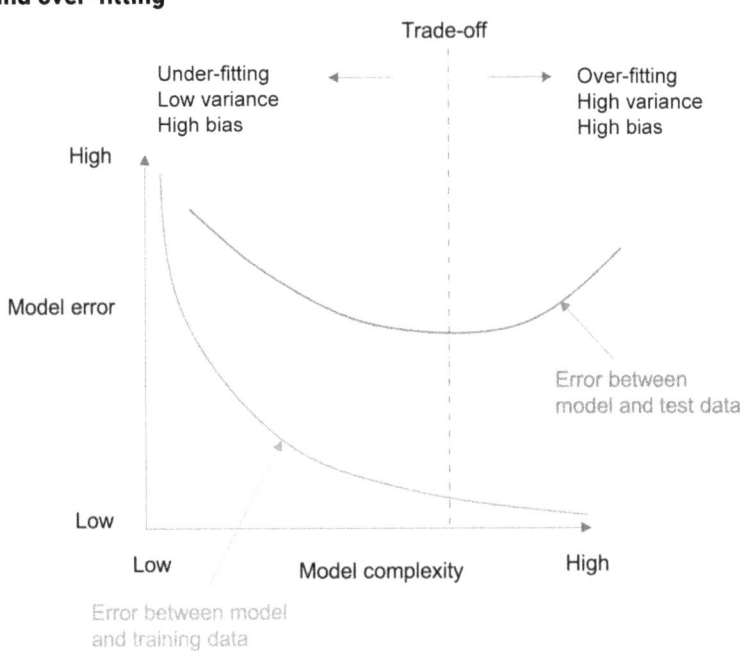

Figure 4.20 Visualisation is learning from experience at every stage of an ML project

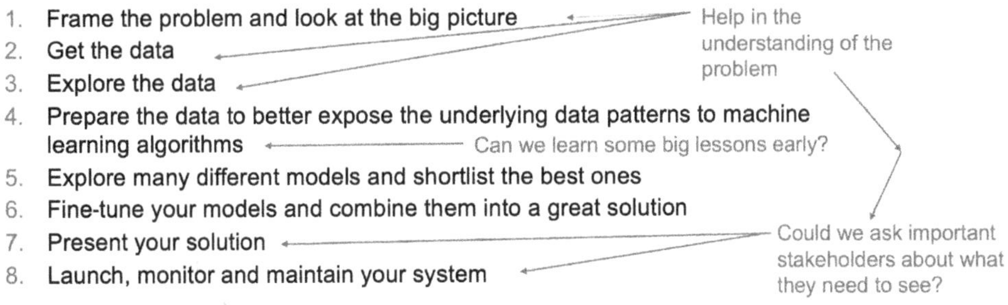

1. Frame the problem and look at the big picture
2. Get the data
3. Explore the data
4. Prepare the data to better expose the underlying data patterns to machine learning algorithms
5. Explore many different models and shortlist the best ones
6. Fine-tune your models and combine them into a great solution
7. Present your solution
8. Launch, monitor and maintain your system
Help in the understanding of the problem
Can we learn some big lessons early?
Could we ask important stakeholders about what they need to see?

It is worth noting that Python is the most common ML coding language of choice for computer and data scientists.[58] The reason for this is often given as the interactive nature of Python. In addition, we often see data pipelines in data visualisation software. Typically, Python is the scripting language within these software packages that automate standard tasks; SciKit-learn and TensorFlow have Python as a coding language.[59] TensorFlow is also widening access to its ML libraries with other languages, such C++ and Swift.

Some typical types of data visualisation are shown in Figure 4.21. The top row are graphs of data that we might encounter in a spreadsheet. The middle row is an iso-contour plot, perhaps of the heights in a map. The image on the right-hand side is an iso-surface; from the iso-contours we can build the 3D iso-surface. With some vector calculus we can also calculate the gradients and perhaps find maximum and minimum heights. These ideas are very powerful if the data we are plotting is the learning rate of a NN or the profit from business activities. We can focus on the fast learning rate or maximise our profit. The bottom row of the examples are networks; these are more complex structures to visualise. Here techniques such as stereo vision, augmented or virtual reality can help. Even large displays such as power walls allow a different perspective on data we are presenting.

These types of data visualisation technique have been around for decades, and software on our computers enable us to create them. However, when we move to large data sets from simulations or Big Data, we need parallel data-processing capability. ParaView (https://www.paraview.org) is a parallel data visualisation ecosystem with Python at its core. It is built on the Visualization Toolkit open-source 3D graphics, image processing and visualisation software library and constructed to run on parallel computers, so can process large data sets and produce graphics that an individual user can see. This is a major challenge when working with large data sets – working with large data sets on parallel computers is not something that should be attempted on your first project.

Most laptop and desktop computers are multi-core and multi-processor machines with graphics processing units (GPUs) that have enough processing power and capability to learn and develop ML/AI. Once tested on small hardware, you can then gradually use greater processing power on high-performance computing systems.

We are fortunate that most open-source software can be compiled, or is already compiled, for common operating systems. You can initially work on your own hardware using the same software you will use on larger hardware set-ups. We are also lucky to have both open-source and commercial software we can draw on. There are advantages and disadvantages to both, and you will need to consider what you will require. A search on your favourite search engine will bring up a list of the many data visualisation packages around.

We know that visualisation is an important aid to learning from experience, but there is an additional idea that we might be able to incorporate: Can we create a learning environment? Can we develop a virtual reality (VR) or augmented reality (AR) simulated environment that we can learn from or humans can teach an AI system to learn from? We mentioned in Chapter 3 that AI robotics gives us the opportunity to explore and utilise extreme environments. Could we build simulated environments to generate the data we need to design and operate our robots? Could we build a learning environment to let a robot learn how to do a task?

Blender is open-source software with a highly sophisticated environment for building simulations and presenting data (see Figure 4.22). The graphics are capable of producing photo-realistic renderings and movies. This is often the base of computer games, movies and scientific visualisation. If we extend this to ML and AI, we could build a human and machine environment where the machine and humans interact to teach each other. Examples of this are found in engineering, where operators are taught how

Figure 4.21 Standard data visualisation examples: (a) line graph, (b) scatter plot, (c) histogram, (d) contour plot, (e) 3D contour plot, (f) network diagram and (g) 3D network diagram

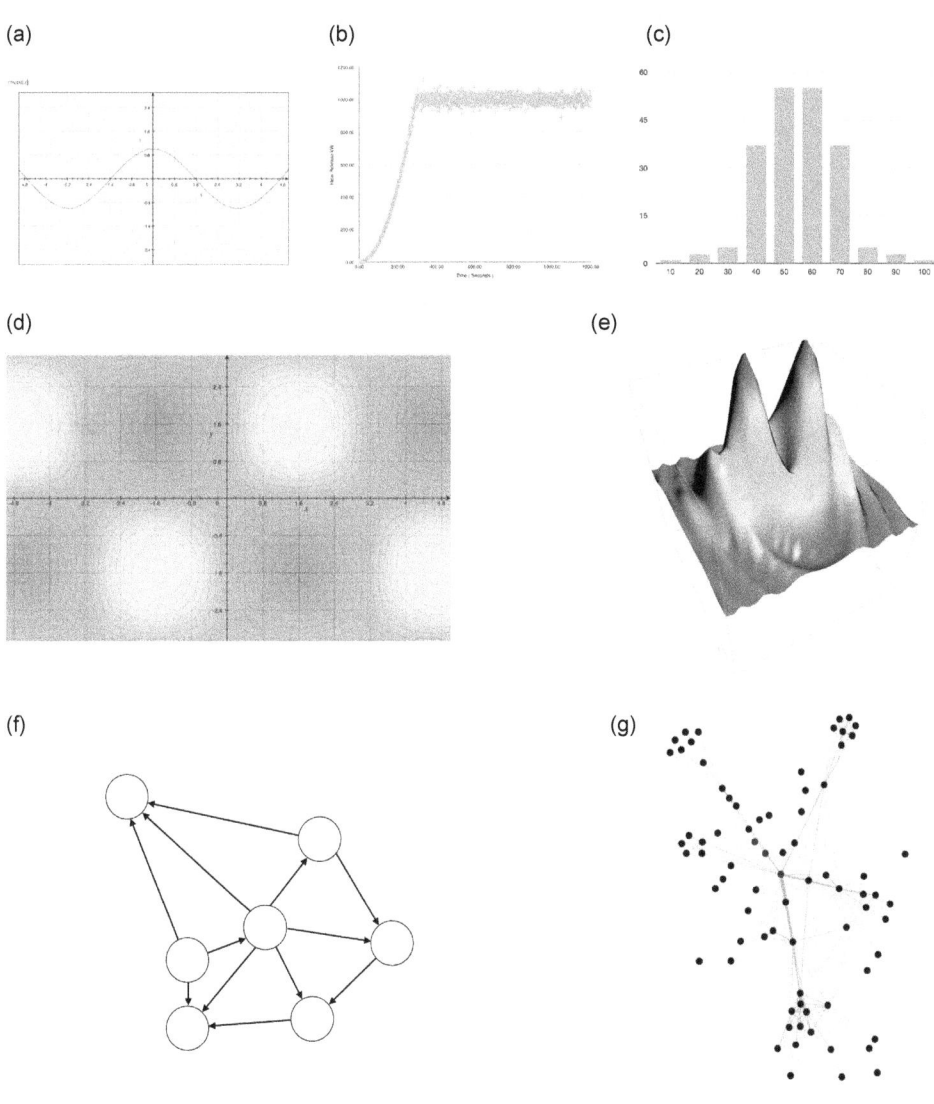

Figure 4.22 Open-source software to build simulated environments – Blender
(reproduced from https://commons.wikimedia.org/wiki/File:Blender_2.90-startup.png under Wikimedia Commons licence)

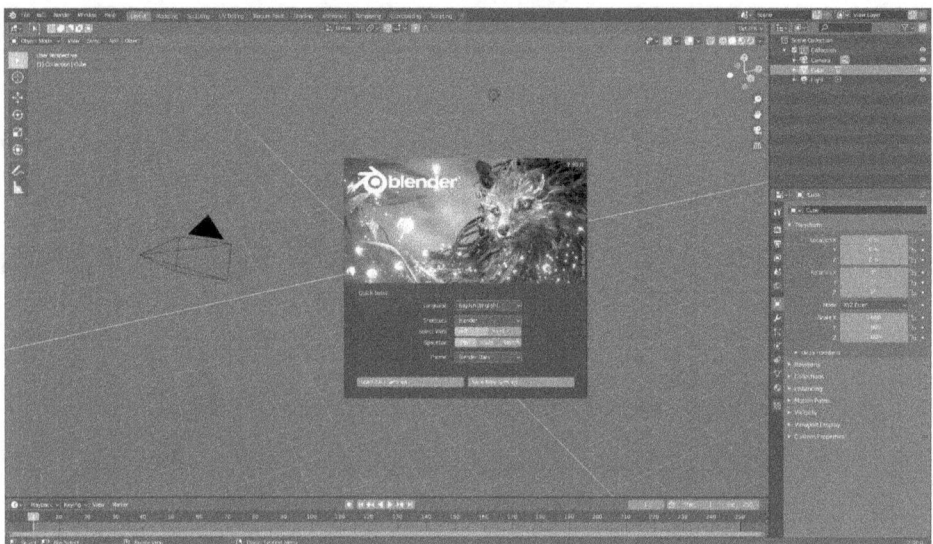

to operate machinery and pilots are taught how to fly in a safe environment. We are always guided by learning from experience, and the learning environment can be used to generate data.

In medicine, AI can be used in diagnosis and treatment. Of particular importance in a surgical environment is what the surgeon sees. This can be developed in a digital environment first, then perhaps a physical environment and finally used on a patient. We can easily imagine a basic system, developing from scanning a patient to one where surgeons can work collaboratively with AI machines. Figure 4.23 shows examples of how AI is working in medicine from diagnosis to treatment. It shows robotic surgical equipment – the surgeon is potentially removed from the sterile operating theatre.

4.2 NEURAL NETWORKS

Here we take a closer at the neural network. In particular we describe the following:

- the link between the NN and the AI agent;
- the basic building blocks of the neural network;
- building algorithms using reinforcement learning;
- a few basic examples of what a NN can do.

Figure 4.23 AI in medicine – humans and machines working together (reproduced from (left) https://en.wikipedia.org/wiki/Radiology#/media/File:Radiologist_interpreting_MRI.jpg; (top right) https://en.wikipedia.org/wiki/Da_Vinci_Surgical_System#/media/File:Cmglee_Cambridge_Science_Festival_2015_da_Vinci.jpg; (bottom right) https://commons.wikimedia.org/wiki/File:Cmglee_Cambridge_Science_Festival_2015_da_Vinci_console.jpg under Wikimedia Commons licence)

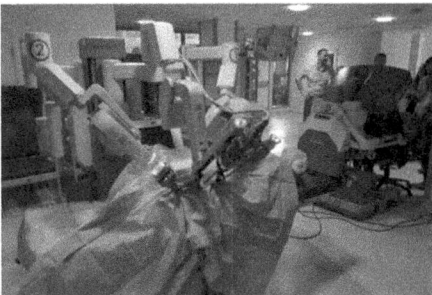
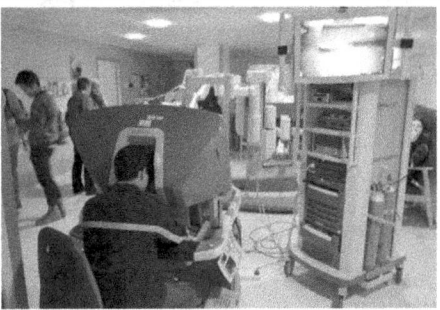

The NN gives machines the ability to **learn from experience** and develop their own algorithms (it is not easy to interpret what these algorithms are, as they are encoded in the NN). In recent years, our understanding on how a NN works has made significant progress. Progress includes winning complex strategy games such as chess and Go, as well as the perhaps more challenging tasks of controlling complex machinery such as robots. In doing so we are giving machines more autonomy.

There are many types of NN available to us, and one that stands out as having made a significant contribution to ML is the convolution NN. However, we are not going to go into detail about this type of NN here because the details are more complicated than the discussion we intend to have here.

To learn from problems that are complex and perhaps random (statistical), we need to use deep neural networks (DNNs). Typically, we employ DNNs for problems such as robot control, image analysis (e.g. medical images) and adversarial game playing. We will introduce what these are and why this makes training them iterative and requires specialist hardware (e.g. high-performance computers, GPUs, etc.) as we progress through the chapter. A DNN has many layers making up the NN; this is why it is called a deep NN.

4.2.1 The link between NNs and the AI agent

The NN is often shown as a sub-topic of ML (see Figure 1.1). The development of the NN started in the 1940s and is based on the biological nature of the human brain. In 1943, Warren McCulloch and Walter Pitts introduced the concept of a neuro-logical network by combining finite state machines, linear threshold decision elements and memory,[60] followed in 1947 by a paper extending these ideas to the recognition of patterns.[61] The first reinforcement NN built by Marvin Minsky occurred in 1951. This was an analogue or electronic device that used reinforcement to learn.[62] This was the start of adaptive control and we see a close link to engineering emerging. In 1962, Rosenblatt named and defined machines called Perceptrons;[63] however, it took until the 1980s for NN to gain significant traction. Today the theory is understood in greater depth. Digital computers have facilitated the use of convolution NN to undertake sophisticated tasks such as learning to control robots and win combinatorically complicated games.

The learning AI agent needs the ability to learn. Reinforcement DNNs or convolution NNs do just that. They can learn from sensors and actuators or from well-engineered computer simulations or games.

Narrow AI has been focused on specific problems. The NN is a more generic learning technique that can be applied to a wider range of scenarios. Training NNs is not easy and understanding their output is difficult also. This poses challenges for justifying them and using them.

4.2.2 The basic building blocks of the neural network

In this section we introduce the concept of learning that is fundamental to NNs. The Perceptron was the initial concept from which our more sophisticated NNs of today evolved. As we build up the basic building blocks of the NN, we will see how the Perceptron fits in.

The NN has a basic mathematical schematic, shown in Figure 4.24, based on the original work of McCulloch and Pitts in 1943.[60] In this figure, we also see a simple representation of a human brain neuron.

Each of these NN neurons form layers (lots of neurons), and multiple (hidden) layers make up a network. The first layer is the input layer where data are fed into the network. The data can be output from a simulation, the colour of pixels in an image or data from sensors. The data pass through the network and as they do, just like the human brain, neurons are activated (or fire). A simplified three-layer version of a NN is shown in Figure 4.25.

There can be any number of layers, and DNNs have lots of layers. Unfortunately, we do not know what number of layers, or depth of DNN, we need before we start training the NN. That is something we learn from experience as the trainer, and part of the hyper-parameters that we will systematically learn from. In this description we have been careful to use only feedforward networks where the input of the node is passed onto the next layer. Recurrent NN also allow nodes to communicate with other nodes and themselves on the same layer. These types of network are described as more dynamic

Figure 4.24 The schematic of a neural network neuron (McCulloch and Pitts 1943) **and a human brain neuron** (reproduced from https://commons.wikimedia.org/wiki/File:Neuron3.png under Wikipedia Commons licence)

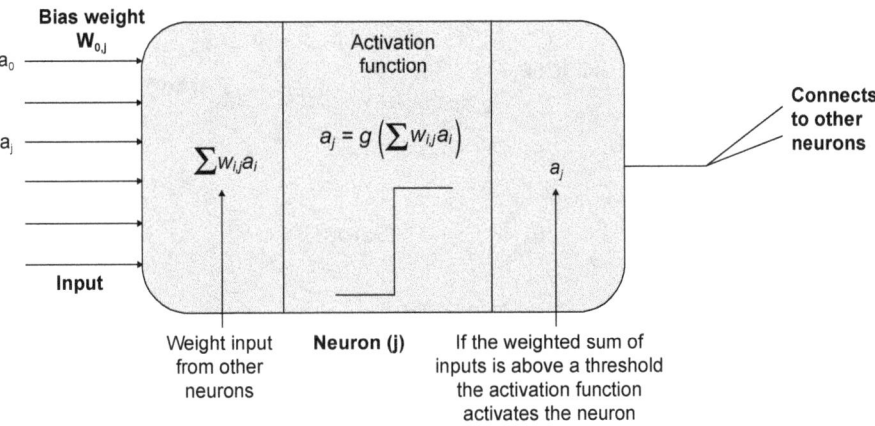

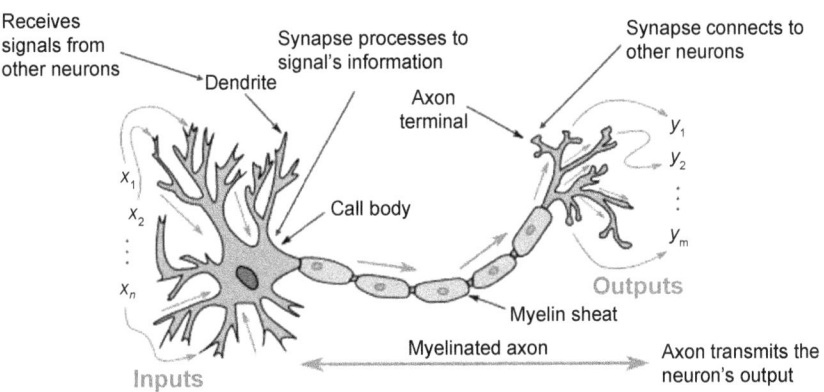

and can behave chaotically. We will stick with the feedforward NNs here, as we are building an intuition or understanding.

NNs work by taking the output (if it is activated) from nodes and passing this information via a link. The link connects the output of a node to nodes on the next layer. Each link from the output of a node is uniquely weighted as it is passed to other nodes. We do not know what these weights are, and training is about determining what the weights should be. An example might help here. Say we have a high resolution picture of a scene taken on your smartphone, the type of picture doesn't matter. We can analyse this picture using a NN. The picture is held on your phone as a matrix of numbers that represent the colour of the image at a particular location. The NN takes this matrix as its input and passes this number to every node on the first layer. Before it does so, it weights it by a unique weight between the input node and the layer node

ARTIFICIAL INTELLIGENCE FOUNDATIONS

Figure 4.25 The schematic of a neural network with an input, hidden and output layer. In NNs, a neuron is a node in the network and the node is made up of an input and activation function and an output. Nodes are connected via links (reproduced from https://commons.wikimedia.org/wiki/File:Colored_neural_network.svg under Wikipedia Commons licence)

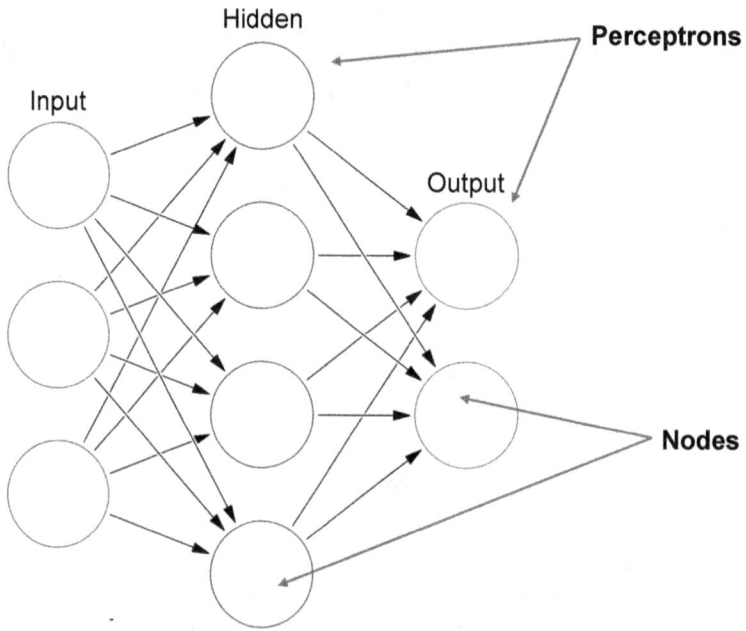

it is going to. Each node on the first layer then adds up the colour number and weights it has received. This sum of weighted input colour numbers is then passed through the activation function. If the output of this operation is greater than a set value, the node is assumed to be activated and the result passed onto the next layer. If it is not activated, then this node is not activated, and its output is zero. The layer of now activated or inactivated nodes is fed into the next layer and the process is repeated. The activation function is trying to represent how our brain fires its neurons when parts of our NN are stimulated. As information is passed through the network of nodes, nodes are fired or not fired. This is a very simple analogy and training a network to find the best weights needs a bias node as well. This helps us to shift the threshold of nodes activating or not.

Whether a node is activated or not is determined by the activation function. The activation function uses the input function as its input. This input function is the sum of the output of nodes connected by the links, multiplied by the weights of the links between the nodes. We then need to determine if the node is activated. If a hard threshold (activation function output is above a set value) is used, then this is a Perceptron. These types of NN can represent non-linear problems that are Boolean in nature. The Sigmoid Perceptron uses a differentiable Sigmoid activation function, and this means DNNs can now represent arbitrary functions, again non-linear. This makes them very powerful. The threshold activation function is a hard step change between the neuron activating or not. The Sigmoid (logistic) activation function is differentiable and is called softer because the decision to fire the node is not a step change. There are other types of activation function, for which use is determined by the experience of the trainer.

The trainer needs to train the neural network to determine the unknown weights of each of the links. This is not a simple task, and a role of the future for AI personnel. Training a NN involves adjusting weights in order for the NN to improve. Techniques involve forward and back propagation to update weights based on the error between the desired output and the NN's output. Knowledge of vector calculus will help here, and a NN domain expert will help too. Up to now we could simplify the NNs as ones that can understand functions, given examples of what those functions are. We could use a NN to learn from examples that we know. This makes the error known. What happens when we do not have labelled examples to learn from? We could build learning examples and try to form labels from our understanding of the NN.

What happens if our problem has lots of unknown combinations – such as a game with many possible moves? We simply may not have the data to understand all the possible moves. We need to measure, somehow, the performance of the NN. This is called reinforcement learning, and the NN can learn by itself. Reinforcement learning fits naturally with the learning agent schematic. The agent has a learning component and the learning component tries to maximise its reward or minimise its loss. Norvig and Russell note that reinforcement learning might be considered as encompassing all of AI.[9] In an environment, an agent will have a percept and will learn via the learning element to achieve a goal. It will use its actuators to act on the environment. This elevates products to intelligent entities and gives products such as robots autonomy. An agent using a NN and reinforcement learning has to learn how to achieve a goal in an environment. This learning is complex. Somehow, we need to train the NN to explore its environment, create a policy to achieve a goal and understand what it can do to the environment with its actuators.

Examples of reinforcement NN successes are:

- AlphaGo, the computer that was taught to beat the world Go champion (see Section 1.3.4). The computer was so successful, it used moves that impressed Go masters. A supercomputer played itself at the game of Go to learn how to play. This is an engineered simulation environment.
- OpenAI, who used an NN to teach a robot how to solve a Rubik's Cube.[64] The robotic hand can manipulate the cube to solve it because it has many sensors and actuators. The NN learned how to solve the Rubik Cube and also how to manipulate its environment.

4.3 AGENTS' FUNCTIONALITY

In this section we give examples of how AI, and in particular ML, is used to give agents functionality. The functionality is used by agents to:

- interpret their precepts;
- undertake an action;
- achieve a goal.

The agent's function is a map of how it understands its percept sequence and turns this into a goal-achieving action. This functionality is usually a high-level or abstract mathematical description of what it is doing. The physical embodiment of the AI agent is then put into the agent program that implements its functionality. Any system could be an agent, certainly from an engineering point of view. Here are some examples of the types of functionality that are useful to an agent:

- decision making;
- planning;
- optimisation;
- searching (the best option or best adversarial game options, route finding);
- natural language processing;
- representing knowledge.

When we think of computational science, engineering, operational research and the statistical and mathematical techniques available, this becomes a very long list of functionality we can build into our agents.

4.4 USING THE CLOUD – CHEAP HIGH-PERFORMANCE COMPUTING FOR AI

In this chapter we have built an understanding of the concepts and theory behind AI using ML as a guide. We can make it complicated if we are training a complex system to achieve a goal; however, once built we can use this learning over and over again. The cloud and the internet make this happen. We now have access to open-source code, improved on an international scale by experts around the world. It is usually free, and the time between academic research and practical application is reducing – progress around the world is accelerating. Software built on object orientation makes code more modular and reusable; libraries of open-source code make development slicker and more reliable. However, there comes a point when we need processing power or the ability to scale applications to a worldwide audience. Here, cloud high-performance computing comes to our aid – it is reliable and cheap (compared to building your own supercomputer).

Typical AI open-source code can be downloaded from a multitude of sources. In some cases, the code will be optimised for use on the cloud service provider's hardware. This isn't always necessary, as you'll be testing and preparing your AI on a laptop or desktop. Cloud service providers also provide online training, forums and user conferences – the cloud community is collaborative.

Data security, as always, is of paramount importance, and ensuring data are protected as the law requires needs careful attention.

You will need to pay for some of the services on the cloud. When starting, there is usually a free period of time given to you for learning. This is very helpful.

4.5 SUMMARY

This chapter has built up an appreciation of the complexity of mathematics and theory that ML and AI bring to our projects. The theory of AI, ML and the subject areas it draws on are vast. With this chapter we hope to have given you an intuitive understanding of what is involved.

5 ALGORITHMS

Algorithms are at the core of problem solving, engineering and computer science. In AI, we are interested in how they can help us with learning from experience; how they can help us to create intelligent entities, products and services that can learn and improve. Algorithms are studied academically, and this can lead to a lifetime's pursuit into complex and rich research. Algorithms are also simple procedures we use to teach ourselves how to program a digital computer (from simple flowcharts to complex event-driven object-orientated software). Narrow AI, and in particular narrow ML, give us a chance to understand how an algorithm can learn from data. The NN and its many variants teach themselves how to solve a problem. It is an algorithm that learns its own algorithm to, say, play chess.

We should note that an AI or ML algorithm may not always be the best option to solve our problems. Scientists, mathematicians and engineers will often have more elegant and efficient ways to solve problems. However, we don't always have this capability, and here ML and AI can help us, especially where we have a lot of data to work with (e.g. data from telescopes, simulations, sensors and so on). Also, algorithms can be expressed in a multitude of ways, as computer code, flowcharts, mathematical notation, pseudocode and so on. For those that want to delve deeper into algorithms, Tom Mitchell's book *Machine Learning*,[23] provides an excellent introduction to many algorithms and their practical uses, including those that we describe below and their derivations.

AI and ML are, at their fundamental core, based on the scientific method, and we must use these techniques with care. Our learning problems are non-linear, iterative and probably statistical in nature. We must use algorithms cautiously because they are often general approaches to problems – domain specific techniques may well be far better suited to our problems. Part of our approach to using ML and AI is to build our learning and experience of what algorithms can achieve.

> ML and AI draw on techniques from well-established academic disciplines such as operational research, control theory, mathematics, numerical analysis and engineering. For example, decision trees are frequently associated with ML, but they have been at the heart of safety engineering and management theory for decades.

In this chapter we will aim to introduce what an algorithm is and why algorithms are so important to ML and AI. Algorithms have their advantages and disadvantages in terms of accuracy, performance and processing time. This chapter includes an explanation and examples of a few of the most common algorithms in ML. We will also explain their use and the challenges that are encountered in deciding which to use and why.

5.1 WHAT IS AN ALGORITHM?

The definition of an algorithm is a process or set of rules to be followed in calculations or other problem-solving operations, especially by a computer. Simply put, an algorithm is a set of instructions designed to perform a specific task. This can be as simple as the process we follow to multiply two numbers together, something that we are taught at an early age and take for granted. Humans do this in decimal, and computers in binary. It is only when you have to explain what you are actually doing that it seems complicated – we need to be explicit because our digital computers need explicit instructions. An algorithm could also be something far more complex, such as compressing a computer file or playing a compressed video file.

We have previously covered learning types and data types, but it is worth addressing the basic concepts again before we go any further into the topic of algorithms.

In a supervised learning model, an algorithm learns on a labelled data set, which provides an answer key that the algorithm can use to evaluate its accuracy on training data.

In an unsupervised model, it uses provided unlabelled data that the algorithm tries to make sense of by extracting features and patterns on its own.

In both semi-supervised and reinforcement learning models, data can be both labelled and unlabelled.

Basic algorithm types include:

- **Regression** – generally used for forecasting and finding out the cause and effect relationship between variables.
- **Classification** – where input data can be separated into groups, for example male or female.
- **Clustering** – the grouping together of data that are similar, for example grouping of images of cats, dogs and rabbits based on data attributes.
- **Association** – where features of data can be associated with other features and, in an unsupervised learning condition, can then associate other data, for example people who generally buy bread and butter also typically buy milk.
- **Control** – where the error measurement between a set point and a measured value will help to decide what to do, for example an autonomous vehicle controller can either apply controlled braking or a collision avoidance manoeuvre.

In Table 5.1 we show a broad correlation between learning models, algorithm types and example algorithms, some of which will be described later in this chapter.

Table 5.1 Examples of the types of learning and algorithms

ML categories	Algorithm types	Example
Supervised learning/labelled data	Regression	Decision tree Linear regression
	Classification	Support vector machine (SVM) K-Nearest Neighbour (k-NN) Naive Bayes
Unsupervised learning/ unlabelled data	Clustering	K-means
	Association	Apriori
Semi-supervised learning/mix of labelled and unlabelled data	Classification	Semi-supervised SVM
	Clustering	K-means
Reinforcement learning	Classification	Ensemble learning – decision forest
	Control	Q-learning

5.2 WHAT IS SPECIAL ABOUT AI/ML ALGORITHMS?

Generally, an algorithm takes some input and uses mathematics and logic to produce the output. In contrast, an AI algorithm takes a combination of both – inputs and outputs – simultaneously in order to 'learn' from the data. In ML, once it has been trained, the ML can produce outputs when given new inputs.

5.3 WHAT ARE SELF-LEARNING ALGORITHMS?

A self-learning algorithm is programmed to refine through iteration its own performance, that is, it learns from itself to improve its accuracy in its ability to perform. In the context of ML, this often requires considerable computational resource. It can be best described as a system into which you feed your requirements (i.e. the desired outcome plus various parameters) and over time the outcome is achieved. An example is recommendation engines where the system gets better and better at recommending 'things', which people then purchase as more data points are processed.

5.4 WHAT ARE THE ALGORITHMS USED IN MACHINE LEARNING AND ARTIFICIAL INTELLIGENCE?

ML and AI algorithms include:

- **Linear regression** – this is a model that assumes a linear relationship between the input variables (x) and the single output variable (y). More specifically, that y can be calculated from a linear combination of the input variables (x). For example, imagine arranging a number of blocks by weight variable y when you cannot weigh them and only know their x (length, height and width) variables. A single input is known as simple linear regression and multiple input is known as multi-variable, multi-variate or multiple linear regression. One use could be to predict a human's ideal weight based on height, build, gender and so on.

- **Logistic regression** – this is a mathematical model used in statistics and ML to estimate the probability of an event occurring having been given some previous data. Logistic regression works with data where either the event variable (y) happens (1), or the event does not happen (0). An example could be around credit card transaction approval when multiple inputs such as time of purchase, place of purchase and type of purchase may determine if the transaction is approved or not.

- **Decision tree** – this is one of the most common ML algorithms in use today. It is a supervised learning algorithm used for classifying problems by moving down a tree root from node to node testing an attribute at each node. It is typically used in expert or smart systems: it can advise a course of action to be taken based on previous successful actions, and is typically seen as a series of questions and suggestions culminating in a 'was this information useful?' question on self-help computer systems. An example could be the fault light on a non-functioning printer: is it red or green? If green, then check if there is paper in the input tray, if the answer is yes, then check for a paper jam and so on, until the problem is identified. The learning element of the algorithm can weight suggestions based on previous successes with particular attributes. Equally, decision trees can be used in association to classify patients and likely disease based on health questions, or loan applications and previous credit history.

- **Random forest** – this is an ensemble learning method for classification, regression and other tasks. It operates by constructing a multitude of decision trees at training time and outputting the class that is the mode of the classes or mean/average prediction of the individual trees. With that said, random forests are a strong modelling technique and much more robust than a single decision tree. Random forests aggregate many decision trees, which limits the over-fitting problem of deep decision trees as well as error due to bias and therefore the ability of the system to give useful and meaningful results. The use of random forest adds additional randomness to the model, while growing the trees. Instead of searching for the most important feature while splitting a node, it searches for the best feature among a random subset of features.

    A commonly used example to explain the difference between decision trees and random forests is that in a decision tree a person may ask one friend to recommend a holiday destination based on a number of questions about preferences and previous holidays, whereas in random forest mode a person may ask the advice separately of multiple friends who again ask a series of different random questions based on their individual likes/dislikes and previous experiences of different holiday experiences.

- **k-NN** – or k-Nearest Neighbour, is a supervised learning algorithm. This means that we train it under supervision and using the labelled data already available to us. It is a relatively simple algorithm used by many organisations and software tools – simple in that it assumes that similar things exist in close proximity to one another. It is typically used in recommendation applications or 'more like this' type systems such as Amazon recommending books or Netflix recommending films based on a particular genre, category, rating, lead actor, director and so on.

The classic travelling salesperson problem is often solved using a nearest neighbour heuristic, and is one of the first algorithms that comes to mind in attempting to solve this particular type of location/route problem. This classic problem is one in which a salesperson has to plan a tour of cities that is of minimal length. In this heuristic, the salesperson starts at some random city and then visits the city nearest to the starting city and so on, only taking care not to visit a city twice. At the end, all cities must be visited and the salesperson must return to the starting city. Use of the k-NN algorithm gives a quick solution compared to a brute force approach where every permutation is calculated. It should be noted, however, that k-NN may not always give the optimal route, especially if there are many data points. The k-NN can be applied to many other movement problems, such as moving a robot or planning the router of a machine tool.

- **SVM** – support vector machine is a supervised ML algorithm that can be used for both classification and regression challenges. An SVM model can be best described as points in space mapped into categories separated by gaps that are as wide as possible. New data points then fall on one side of the gap and are then placed into that category. SVMs have many uses ranging from image recognition to handwriting recognition to satellite data classification. A simple example may be categorising images of cats and dogs where an image must fall on one side or other of the gap.

When supervised learning is not possible due to unlabelled data, then an unsupervised approach needs to be taken. SVM attempts to find natural clustering of data into groups and new data will then fall into one of these groups.

- **Naive Bayes** – is a classification technique based on Bayes' theorem with an assumption of strong independence among predictors. Simply put, a Naive Bayes classifier assumes that the presence of a particular feature in a class is unrelated to the presence of any other feature. It is used in predicting membership probabilities for each class, such as the probability that a given record or data point belongs to a particular class. A simple example could be the prediction of the probability of you having a particular illness or disease based on the data recorded about you, which may include ethnicity, age and gender, and not just a list of dependent symptoms such as a rash or high temperature.

- **K-means** – this is a centroid based clustering algorithm, which means that data are clustered around a centre. K-means is an iterative algorithm, and has two unique steps: the first being a cluster assignment and the second being a move to the centroid step.

Initially, you must select a number of clustered centres depending on the number of clustered groups you want to create. Measurements are then made from the various data points through each of the data points and, depending on which cluster is closer – whether the cluster A centroid or cluster B centroid or cluster C centroid and so on – the algorithm assigns the data points to one of the cluster centroids. K-means then moves the centroids to the average of the points in a cluster. In other words, the algorithm calculates the average of all the points in a cluster and moves the centroid to that average location. The two steps are then repeated until an end condition is met.

A real-world example of this algorithm is in the segmentation of customers for marketing purposes into various persona categories to allow better targeted marketing messages to be sent to the different personas, thereby helping to increase engagement or sales. In marketing you would typically try to limit the number of personas to a manageable number – let's say six would be optimum – but how we determine that for our particular industry or market segment requires some learning from experience.

The choice of the number of clusters to determine the optimum number of clusters (i.e. the 'k') can be quite complex and may be decided by adding an additional cluster until it no longer makes a significant difference. A single data point could also be considered a cluster. The decision criteria for k is beyond the scope of this book. Again, we see the importance of learning from experience while we are building our AI or ML technique.

The examples above are the tip of the iceberg in terms of the algorithms used in ML and AI, where there are literally thousands, if not millions, used for specific industries, for solving specific problems or within individual applications.

5.5 WHAT IS A DEEP LEARNING ALGORITHM?

As with the algorithms described in the last section, there are also a number of algorithms used in deep learning on neural networks. The detailed workings of the algorithms are beyond the scope of this book other than to provide a description of their function and use. These algorithms include:

- **Multilayer Perceptron neural network (MLP NN)** – is a class of feedforward artificial neural network (ANN). MLP utilises a supervised learning technique called backpropagation for training. Its multiple layers and non-linear activation distinguish MLP from a linear Perceptron. It can distinguish data that are not linearly separable.

- **Backpropagation** – short for 'backward propagation of errors', is an algorithm for supervised learning of ANNs using gradient descent. Given an artificial NN and an error function, the method calculates the gradient of the error function with respect to the NN's weights. Weights can then be adjusted to improve the error.

- **Convolutional Neural Network (CNN)** – is a deep learning algorithm that can take in an input image and assign importance, based on learnable weights and biases, to various aspects/objects in the image and be able to differentiate one from the other. The CNN was inspired by the human brain's processing of visual data. These NN are more versatile and can be used on data other than images (e.g. audio, self-driving cars, and more).

- **Recurrent neural network (RNN)** – is a class of ANN that processes a sequence of data along a timeline and remembers 'in memory' the sequence of data or a variable amount of data to use later in its processing. There is a connection between the data points, just as we have when we read individual words and link them together in a sentence. RNNs have been derived from feedforward NNs. They are typically used in speech recognition and connected handwriting recognition applications where what has come previously may influence what comes next or how it is interpreted.

- **Long short-term memory (LSTM)** – is an artificial RNN architecture used in the field of deep learning. Unlike standard feedforward NNs, LSTM has feedback connections. It can not only process single data points (such as images), but also entire sequences of data (such as speech or video).

- **Generative adversarial network (GAN)** – is a class of ML frameworks designed by Ian Goodfellow and his colleagues in 2014: two NNs contest with each other in the form of a zero-sum game, where one agent's gain is another agent's loss. Apart from gaming, GANs are also used in image improvement and event generation and are used in the generation of 'deepfake' photorealistic human images resembling real photographs and videos generated from a single image.

- **Restricted Boltzmann machine (RBM)** – is a restricted two-layer ANN with a visible and a hidden layer. Originally invented in the 1980s, and still used today, it can learn probability distribution from its data inputs. RBMs can also be trained in either supervised or unsupervised ways, depending on the task. RBMs are used in topic modelling, classification, collaborative filtering and feature learning, and have found uses in other applications such as speech recognition. The RBM is the basis for DBN.

- **Deep belief network (DBN)** – can be created from multiple layers of simple, unsupervised networks such as RBMs. The DBN in its simplest form is composed of many layers of latent variables, with connections between the layers but not between units within each layer. When the DBN is initially trained without supervision on a data set it can learn to construct its inputs probabilistically. After initial training, supervised training will then allow it to perform classification. DBNs are often used in the field of drug discovery.

Many other deep learning algorithms are available, but the descriptions above show that significant skills, knowledge and experience are required to choose, apply and make successful use of these algorithms. Each type of algorithm described above is a rich area of mathematics and theory. They are all rooted in the scientific method, and a rigorous approach to the use of these algorithms is likely to be found. What we mean here is that, although the interpretation of a complex algorithm is difficult, its roots in the scientific method will help to build a professional and rigorous justification for their use and results. Deep learning approaches contain more data and steps than a human is capable of comprehending. Again, we should remind ourselves that these general learning techniques may not be the optimum solution to our problems. Here, the domain expert will play a pivotal role in ascertaining what is fit for purpose. They will also be able to help with heuristics that are so vital to ML.

5.6 THE ETHICAL USE OF ALGORITHMS

The ethical use of algorithms is a topic growing in importance as the use of AI systems becomes more prevalent in society.

In March 2016 Microsoft's Twitter chatbot 'Tay', which was targeting 18–24-year-old millennials, started spouting racist, anti-feminist and pro-Trump tweets, resulting in Microsoft having to remove the bot less than 24 hours later.[65] In its defence it was merely doing what it had learned from other Twitter users. It may be a surprise to some to learn that algorithms themselves are not biased; the bias comes from the data that are used to train the ML system. In this case, it included many racist and misogynistic tweets. Microsoft said that it would revive Tay only if its engineers could find a way to prevent web users from influencing the chatbot in ways that undermine the company's principles and values.

In 2020, because of COVID-19 in the UK, school closures caused by the national lockdown resulted in children being unable to sit their GCSE exams. The Department of Education decided to use an algorithm to predict GCSE exam results rather than use teacher-predicted grades. The decision hit the news when tens of thousands of children received significantly lower marks than their teachers had predicted. This led to widespread complaints and protests, and resulted in the government backtracking on their decision to use an algorithm and upgrading many of the exam results to the teacher-predicted marks. A lot of bad press followed in the British media and caused many people to believe that algorithms cannot be trusted.

Whether you are aware of them or not, algorithms are becoming an everyday part of our lives and are used in almost all industries and supporting technologies that we take for granted and use on a daily basis, so the ethical use of ML and AI is more than just a focus on algorithms, it must include an ethical purpose. See Section 1.3.10 for the discussion on AI ethics.

5.7 SUMMARY

Algorithms are the steps that underpin AI and ML. The use of AI-enabled products and services within society impacts our everyday lives. Some of them are highly dependent on the quality of input data or data used to train them, others can generate their own data.

This chapter has given you an insight into algorithms and the understanding that they require careful consideration regarding their selection and purpose.

6 THE MANAGEMENT, ROLES AND RESPONSIBILITIES OF HUMANS AND MACHINES

This chapter introduces the concept of humans and machines working together. What does it mean for humans and what does it mean for machines? It covers the types of role there will be as we move forward and, as learning from experience is at the core of AI, what project style will most likely help you to achieve your projects' goals.

6.1 ARTIFICIAL INTELLIGENCE WILL DRIVE HUMANS AND MACHINES TO WORK TOGETHER

Ever since the realisation that artificial intelligence was possible, we have considered the pros and cons of exploiting AI. It is only in the last couple of decades, when it has become technologically possible on a large scale, that we have seriously considered the challenges, issues and ethics around it; in particular, intelligent learning machines replacing humans in various roles. In fact, in some industries, automation has been a challenge since the first industrial revolution. In other industries and professions, it will be a long time until a machine will replace a human and, in certain roles, it will likely never be the case.

It is likely that in many aspects of our personal and working lives we will not be replaced by AI, but work alongside learning machines in some way.

6.1.1 Augmenting humans and machines

If we take the next step from ML to thinking about humans and machines together, we now need to consider how humans can augment machines and how machines can augment humans. We are at a point where the ethical purpose of our AI system is important. Roles for humans and machines will certainly emerge to ensure our AI systems have a human-centric ethical purpose.

6.1.1.1 Augmenting humans
Firstly, let's look at ways in which machines can augment human capability through the effective use of information technology. Improvements can range from slight increases in performance right up to making us super-human.

To do this we need to introduce a new concept: intelligent automation or augmentation (IA). IA aims to use ML technologies to assist rather than replace humans. IA can take technology such as computer vision, NLP and ML and apply it to RPA, allowing the automation of processes that don't necessarily have a rules-based structure, for example using an IA system alongside an oncologist to identify cancer cells in biopsy slides – previously the task would have required two oncologists. This is an example of human and machine working alongside each other for the benefit of humankind.

Recent advances in IA technologies have helped to advance human potential by increasing worker productivity, alleviating mundane and repetitive tasks and introducing and enhancing convenience in our lives. For decades, some machines have been much better than humans at performing certain tasks, but what we increasingly see now is heightened penetration and a continuous acceleration of IA across many industries and spheres of life. Over the next decade, we will see further rapid advances along these fronts as technology improves and acceptance of IA increases.

6.1.1.2 Augmenting machines

The key areas where humans augment machines are in making AI and IA more understandable and acceptable to people in general.

At a simple level, machines can increase our learning and make it more effective. We humans have to orchestrate that learning and provide support when the machine hits a contradiction or fails. We also need to ensure that machines are doing the right things. Remember, we need human-centric ethical purpose to our AI systems. Machines don't have consciousness or emotion, which makes empathy a big problem. Humans will need to set the goals not just about where we are going but also how we are going to get there.

6.1.2 The AI ethicists, leaders, testers and trainers

In this section we describe some of the roles that have evolved and are emerging as we develop AI systems and services. Some of the roles are still evolving and not yet fully formed, so here we will talk in generic skill terms; later we will focus on specific roles within an AI project.

Firstly, let's talk about why we need different roles such as AI ethicist, leader, tester and trainer. Documented below are commonly advertised roles.

- **AI visionary/lead/decision maker –** you will certainly need someone within your organisation with a vision of what they want to achieve with AI or how they want to exploit the potential of AI. This person will be the driver for change, a visionary, even an entrepreneur like Elon Musk.

- **Data scientist –** you will need someone who understands how to extract, analyse and interpret large amounts of data from a range of sources using algorithmic, data mining, artificial intelligence, machine learning and statistical tools, in order to make it accessible to businesses. Data scientists analyse data for actionable insights, helping to devise and apply models and algorithms to mine the stores of Big Data and analyse the data to identify patterns and trends.

- **AI tester/AI test analyst –** this is a test management expert who specialises and focuses on testing the AI system/product throughout the software development life cycle and implementation life cycle, typically working alongside data analysts.

- **Algorithm specialist** – you will need someone, most likely a computer scientist, who can perform research and design algorithms for real-world applications. Algorithm specialists perform research to learn how to make faster and better-running sequences for applications that are often too complicated or not yielding acceptable results.

- **AI oversight manager** – you will need someone to provide oversight of an AI project, someone who can spot things such as data bias. Specialists will be needed to interpret AI's decisions and make sure they are fair and transparent.

- **Applied machine learning engineer** – an applied AI/ML engineer's best attribute is not an understanding of how algorithms work; rather, their job is to use them, not build them (that's what researchers do). What you are looking for in ML engineers are experts at coding that get existing algorithms to accept and churn through your data sets. They drive the ML to work well.

- **AI ethicist/data ethicist** – if we develop an autonomous car that causes a crash, who is to blame if all the decisions made by its algorithms have first been programmed by humans? AI-powered platforms are likely to throw up a host of ethical dilemmas, and organisations will require hordes of AI and data ethicists to help address them.

- **Social physicist/scientist** – our increasingly advanced AI tools will enable researchers and governments to process vast amounts of both structured and unstructured data and find insights that help to solve real-world problems, be it tackling global health epidemics, famines or crime. Processing Big Data will be key to effective solutions. During the COVID-19 crisis in 2020, epidemiologists used various algorithms modelling masses of data to identify transmission mechanisms and transmission rates.

- **Expert analyst/analytics manager/data** – this role tends to be a bit of a hybrid one; a hybrid between the data scientist and the decision maker. The role is to keep the other data scientists on track and focused on achieving the project objectives.

- **AI trainer** – this role is wide ranging, but is primarily to make sure that data are properly annotated, within extremely strict annotation guidelines, to achieve required outcomes from the AI system. This may be to help natural language processors and language translators make fewer errors or to train AI algorithms how to mimic various human behaviours.

- **Robotics/cobotics trainer** – this role ranges from training basic robots to highly collaborative AI-powered robots on production lines or those that are autonomously working. Humans are increasingly likely to work alongside and collaborate with robots on many tasks and processes. Knowledge of how to train deployed robotic units, and knowing how to control these machines, will be vital. Training robots and humans to work collaboratively will ensure they work efficiently together.

- **AI explainer** – these typically work directly with management and leadership, bridging the gap between technologists and management to explain what the AI system is actually doing and explain to some extent the nature of what is often seen as 'black box' technology. This may involve testing, observing and explaining the algorithms, and improving the overall explainability of the system and its interfaces, even to the point of justifying decisions an AI system may make. The role often incorporates the necessity to effectively communicate the benefits of AI and changes throughout the organisation and maybe to the wider public. The role may also include showcasing the future work of AI and ML to customers, NGOs, government and wider society to gain acceptance of AI-powered systems and services.

- **Technology convergence/integration technicians** – as we will most likely not be using our AI service in isolation, but alongside a host of other disruptive technologies, from VR and 3D printing to the IoT, we will probably require specialists who can help organisations integrate these technologies or combine them to create new AI offerings.
- **AI sustainers** – this role is based on maintaining and sustaining deployed AI systems to ensure continued confidence is maintained. This will involve managing and optimising the daily operations of the AI systems, robots and/or software post-implementation – installing 'guardrails' to alert people when the AI system stops performing as expected.

6.2 FUTURE DIRECTIONS OF HUMANS AND MACHINES WORKING TOGETHER

Could you jump backwards 20, 30 or 40 years and foresee what you are doing today? Could you have predicted the technologies and systems we are using now? I suspect not. For some, the world we live in today is a disappointment as we were, arguably, promised an abundance of personal robots, flying cars and affordable spaceflight. We are still waiting for that to become commonplace although, for what it's worth, we do have a couple of autonomous rovers on the planet Mars.

Forecasting where we will be in 2030, 2040 or even 2050 is equally as difficult, but we try based on current research and technology innovations that are in the pipeline. If you want to see different possible outcomes, Max Tegmark's book, *Life 3.0*,[34] is an intelligent and perceptive exploration of possible AI futures, and Ray Kurzweil's book, *The Singularity is Near*,[10] is an optimistic look at the future.

6.3 'LEARNING FROM EXPERIENCE' AGILE APPROACH TO PROJECTS

Can you conduct an AI project using a waterfall methodology? Of course you can, as long as you understand the end-game before you kickstart the project. Unfortunately, many AI projects don't know the end point from the outset, so naturally lend themselves towards a more Agile approach to project management, which incorporates a certain amount of experimentation.

Who should be part of an Agile AI project team?

6.3.1 The team members needed for an Agile AI project team

Who you will need on your project team depends on the type and the scale of the project. Are you just bolting an AI application to an existing service? Are you replacing a legacy service with an off-the-shelf AI product? Is this your organisation's first foray into AI and will it involve a lot of discovery and experimentation that may have little or no residual business value? Or is this a bespoke AI solution you are developing in-house or alongside a third-party organisation? Regardless, there are certain roles that you cannot do without.

You will certainly need an AI product owner/product manager, someone who both understands the consumers' requirements and has the ability to control the content of various development sprints and control the overall direction that the project is heading in. AI product management is focused on using AI, deep learning and/or ML to enhance, improve, create and shape products. This is also your AI decision maker.

Organisations increasingly realise that AI is fast proving a business differentiator, and project teams are rapidly springing up, leading to a shortage of skilled and experienced professionals. Here are the key roles and skills you may soon need to fill for your AI project team:

- **Domain expert** – this is your subject matter expert, and their skills will depend on the nature, size and complexity of the project. If the project is to develop a AI-driven medical app, then this person is likely to be someone with specific medical expertise in the area you are working on. If the project has a legal basis, then the domain expert is likely a lawyer. You may also require more than one domain expert.
- **Ethicist** – this is the person that advises the team on ethical AI practice, guards against bias and unintended consequences and ensures accountability within the organisation. They also hold a mirror to what the culture and principles say and what the real practices do. This is the person you go to if something 'feels wrong or ambiguous'.
- **Software engineer(s)** – these are the computer science professionals who use their knowledge of engineering principles and programming languages to build the AI software and hardware products. These are your workhorses.
- **Reliability engineer** – this is the person who will quality assure the project by focusing on areas such as loss elimination, risk management and life cycle asset management.
- **User experience (UX) designer** – this is the person who will ensure that user requirements have been built into the design and test criteria by conducting user research, designing, writing UX copy, validating/testing with users and selling/presenting the design solution to the business. It is the UX designer's role to be the voice of the user and advocate for the user's needs while balancing the business goals. For instance, if our project is an autonomous vehicle, then this role may well include an automotive anthropologist specialist.
- **Interactive visualiser/graphic designer** – this is the person who establishes the look and feel for the various interfaces, including websites, mobile devices, apps, kiosks, games and wearables. They work within brand guidelines to create layouts that reinforce a brand's style or voice through its visual touchpoints.

- **Data collection specialist** – this is the person who develops and maintains collection plans and protocols for data to support a specific AI-based product or service. They coordinate field collection activities, including logistics (locations, equipment, personnel), to track, organise and curate data collected in the field. If it is an agricultural AI project, for instance, they may literally be out in a field.
- **Product owner/project/programme manager** – this is the person/people who know how to run a project, and keep the project focused, on track and within budget.

Each of the roles/competencies defined above will add to the success of the AI project, but each AI project is different and unfortunately there is no one-size-fits-all team format; you'll have to work it out yourself. The Skills for the Information Age (SFIA) Foundation framework is, at the time of writing, in the process of being updated to incorporate details of the skills and competencies we have outlined above (https://sfia-online.org/en).

6.4 SUMMARY

This chapter has outlined the future of humans and machines working together. It gave examples of the roles machines and humans will likely undertake, and the type of project style those using AI might adopt.

7   AI IN USE IN INDUSTRY: REIMAGINING EVERYTHING IN THE FOURTH INDUSTRIAL REVOLUTION

We have previously discussed what the fourth industrial revolution is and how, arguably, we are still in the middle of it and contributing to its progress. We therefore have an opportunity to look at what we do and how we do it to see if it can be done better, cheaper, quicker and more efficiently, improving all our lives and those of future generations without impacting on us achieving our common and shared sustainability goals.

7.1 AI ALREADY IN USE

Artificial Intelligence has been in use in various industries for many years, although some sectors are only just now discovering the benefits and the potential challenges that AI brings to their industries and organisations. Here we will touch on where we stand at the time of writing (2020), what has been achieved in terms of AI use cases and what is being developed for the future.

7.1.1 Research and development (R&D)

R&D is the professional embodiment of learning from experience and the scientific method, and therefore AI is a fundamental part of R&D. Both AI and ML are now considered to be part of the modern R&D toolkit for the innovation engine room. Use of AI can drastically reduce R&D timescales and assist in product development by allowing designers to explore a wider range of design options than can be created by the human mind alone.

> Occasionally AI systems have come up with designs that cannot be manufactured, in much the same way that AI systems for music composition have developed piano scores that cannot physically be played by a human pianist.

Common use cases in R&D include:

- **Modelling** – researchers have used AI to model scenarios with experimental selection before building a physical experiment. By using AI it is possible to simplify analysis by coupling sensors to simulations, for example modelling the mechanics and practicalities of an automated picking and packing warehouse.

- **Analysis** – weather analysis has also used AI in ensemble simulations analyses using ML 'most likely' forecast.
- **Testing** – the application of AI in software testing tools is focused on making the software development life cycle easier. Through the application of reasoning, problem solving and, in some cases, ML, AI can be used to help automate and reduce the amount of mundane and tedious tasks in systems development and testing.
- **Automated research** – researchers have also used AI in literature searches to identify key concepts and papers and collect data, thus saving researchers' time and allowing them to focus on the hypothesis and results rather than data collection. AI also enhances human capabilities in areas such as data mining and analytics.

7.1.2 Health and social care

Increasingly we see that health and social care are inexorably linked to human wellbeing throughout our life. There is a huge potential for the use of AI-powered services for the betterment of mankind. Specifically, it increases the ability for healthcare professionals to better understand our day-to-day patterns of activity, lifestyle choices and needs; with increased understanding, they can provide better treatments, healthcare guidance and proactive advice. AI is transforming all aspects of healthcare – diagnoses, treatment and health and care delivery – to be more efficient and cost-effective.[66]

There is also potential for learning and further AI application in social care settings based on data collected through new technologies such as smart sensors in homes and telemedicine robots. However, the potential benefits of increased use of AI need to be carefully weighed against privacy and safeguarding implications now and in the future.

Key areas of the current and future use of AI in healthcare are:

- **Diagnosis and treatment selection** – according to the American Cancer Society,[67] a high proportion of mammograms yield false results, leading to one in two healthy women being told they have cancer. The use of AI is enabling review and translation of mammograms 30 times faster with 99 per cent accuracy, reducing the need for unnecessary biopsies.
- **Drug development** – AI has the ability to analyse big data sets, pulling together patient insights and leading to predictive analysis. It can also help to understand disease mechanisms, identify drug targets, find good molecules from existing data libraries, identify personalised/precision medicine, suggest chemical modifications and identify candidate drugs for repurposing.
- **Smart hospitals and improved patient interaction** – AI-powered digital transformation has the potential to create smart hospitals and improve patient interaction at each contact point from appointment management through to discharge from treatment. It could have the greatest potential for positive impact on our often-overstretched healthcare infrastructure.

- **Patient monitoring** – healthcare, as we know, produces vast amounts of data and, as we also know, data are literally the lifeblood of AI. Wearable and implanted biosensors will allow for real-time patient monitoring while in hospital and in the comfort of the patient's home. There is also the potential to use AI and Big Data processing techniques to provide better data and analytics to clinicians, enabling them to deliver better care.

- **Robotic surgery** – AI has extended a surgeon's capability even to the nanoscale. Surgical robots have been approved in the USA since 2000, and provide surgeons with superpowers – improving a surgeon's ability to see, create precise and minimally invasive incisions and also stitch wounds. The robots can take a lesser role surgically, allowing the surgeons to make the most difficult and important decisions.

- **Other robot uses** – other uses include room disinfection using ultraviolet (UV) light, reducing the chances of a hospital-acquired infection such as Methicillin-resistant *Staphylococcus aureus* (MRSA) or *Clostridioides difficile* (C. diff). On a reception desk, a robot could register patients in several languages. Robots can carry supplies from one department to another; be programmed to respond to shortages; even deliver food to patient rooms. Some robots can assist patients getting in and out of beds or wheelchairs, or improve mobility for patients with partial paralysis. Newly developed robots can assist in drawing blood by pinpointing the ideal vein and withdrawing blood in half the time it takes a nurse to do the same thing.

- **Image analysis** – image recognition is fundamental to AI and, when linked with analysis, opens up a number of areas to help detect malignancies, reduce occurrence of false positives and detect instances when these may have been missed when relying on purely human observation – leading to more improved outcomes. AI-powered imaging of X-rays, computerised tomography (CT) and magnetic resonance imaging (MRI) scans have also extended image analysis into the areas of fracture detection, cardiovascular abnormality identification and neurological disease detection.

- **Genetic analysis/DNA** – AI tools can support healthcare professionals to provide a faster service, diagnose issues and analyse data to identify trends or genetic information that would predispose someone to a particular disease. This allows early intervention or even prevention.

- **Patient engagement** – AI can be used to engage with patients to assist and support them with pre- or post-operation activities such as weight loss prior to surgery, for instance, or even help soothe the nerves of nervous or traumatised patients.

- **Stroke rehabilitation** – stroke patients and others with neurological injuries are turning to the rehabilitation powers of robots with AI that help people regain previously lost mobility. Various studies have demonstrated that by using robotic therapy, patients can heal faster and better than with any conventional therapy.

In social care settings, AI is being used and being developed for:

- **Monitoring and support at home** – AI can help to support assisted living through the development of 'smart home' AI technologies, including movement and/or sound detection technologies. This could indicate to a carer that a person in their care has got out of bed or had a fall or isn't responding to a doorbell, for instance. This can provide peace of mind to carers and relatives when managing someone, for example, with early stages of dementia living in a private home setting.

- **Robotics** – in a homecare environment, robots can assist with movement, bathing, food preparation, housework, toileting and so on. In a study in Boston, Massachusetts, in 2011, robots were sent home with newly discharged children to monitor their medical condition.[68]
- **Social interaction** – AI can be used to increase social interaction for people living on their own using various technologies, such as personal assistants, avatars and so on, to improve communication and help with increased independence.
- **Home automation** – increased use of AI in home automation and increased accessibility through phone apps are allowing people who would previously have needed to live in care homes to live increasingly independent lives.

See Todd Stabelfeldt's film, *Convenience for You is Independence for Me*, at https://developer.apple.com/videos/play/wwdc2017/110/.

7.1.3 Finance services

The financial services sector is one of the major users of AI and technology that enables near-human levels of cognition. AI shows great promise for the financial services industry, which is using it in many different ways including:

- **Improving consumer products and services** – AI is transforming the way services are delivered to customers, allowing more intelligent and tailored products and services to be presented to customers who increasingly want more-convenient, safer and smarter ways to access, save, invest and spend their money.
- **Cybersecurity** – AI is enhancing cybersecurity, helping banks by more easily identifying consumer fraud, detecting anti-money laundering patterns and making recommendations to customers to reduce current and future security risks.
- **Business process efficiencies** – within both capital and investment markets, AI can help to produce better, faster and more accurate predictions of future financial outcomes and trends. With its power to predict future scenarios by analysing past behaviours, AI is helping banks and, increasingly, individual traders to predict future outcomes and trends.
- **Credit checking** – AI systems can analyse thousands of data points from various credit bureau sources to assess credit risk for small business and consumer loan applicants. This utilises ML to find patterns and determine good and bad applications.
- **Personalised banking** – many institutions are making use of AI assistants, such as chatbots, and use AI technologies to generate personalised financial advice and NLP to provide instant, self-help customer service.

7.1.4 Education

While AI can teach students new skills or reinforce difficult concepts for struggling students, it cannot yet replace a human teacher. AI programs have proven that they can teach students to read or do logic/mathematics, but as yet have failed in the teaching of social and emotional skills, which are more complex.

The use of AI across education includes:

- **Augmenting the teacher** – AI can help to teach students new skills or reinforce difficult concepts for struggling students, effectively providing an extra set of hands, eyes and ears.

- **Automating administration** – AI can automate the administrative duties for teachers and academic institutions alike. A lot of time spent marking exams, assessing homework and providing feedback to students can be automated, increasing the time teachers have available to interact in more beneficial ways with their students, using their unique human capabilities where machines would struggle.

- **Improving exam performance** – AI helps by improving exam performance. Each time a test exam is taken the information gained on previous performance can be used to give feedback on the areas that need to be worked on and improved, and identify areas were performance is satisfactory.

- **Individualising intelligent tutoring systems/personalised learning** – AI has the potential to deliver truly personal learning programmes for students where it has not been possible in the past. This could replace the current situation of each class following the same learning programme, often at the same pace, meaning some students are bored and some left behind. Several colleges and universities are using cognitive science and AI technologies to provide personalised tutoring and real-time feedback for post-secondary education students. It will only be a matter of time before this is more widely adopted across all educational settings.

- **Delivering smart content** – AI can be used to help disseminate and break down textbook content into digestible 'smart' study guides that include chapter summaries, true–false and multiple-choice practice tests and flashcards.

- **Reducing administration tasks** – AI will take some tasks off teachers' hands, allowing them to focus elsewhere. With AI grading students, developing a personalised curriculum, identifying gaps, creating smart content and making education more accessible, teachers will become coaches or facilitators. AI has already automated drab administrative tasks and cut down on the time and effort teachers used to put into activities such as assessing students' work and grading exams.

7.1.5 Logistics – planning and organisation

Planning the movement of millions of items is heavily dependent upon statistical data, which are core to inventory management. In our global economy, an understanding of the many AI approaches is important for executives trying to implement innovative AI-driven software in their logistics companies to reduce the cost of the worldwide logistics challenge.

Some example uses of AI in logistics include:

- **Automated planning** – AI allows logistics companies to more accurately predict the movement of products and goods. It does this by correcting data quality problems at an early stage and planning the movement logistics of data sets of disparate goods, some of which may be time or heat sensitive. This is the basis of 'smart integrated delivery'.

- **Supply chain management** – AI provides the supply chain with contextual intelligence that can be used to manage inventory and operating costs more effectively, creating 'smart integrated supply chains'. Use of technologies such as Blockchain in key supply chain activities will help to improve supply chain efficiency, reliability, security and integrity. Technologies such as vehicle telematics can also assist in operational management through the use of global positioning system (GPS) enabled vehicle operational data, currently being used in delivery fleets equipped with electronic logging devices that can be tracked by their fleet operators in real time. Imagine coordinating a delivery fleet of electronic vehicles that need to be charged at selected locations.

- **Intelligent and automated warehouse management** – companies use AI with ML to get new insights into warehouse management, logistics and supply chain management, helping to develop 'smart warehouses'. They use technologies such as autonomous robotics, intelligent agent modelling (learning from simulated experience) and knowledge engineering (learning from the experience of experts). Organisations, such as Ocado in the UK, minimise floor space in their warehousing by using robots that can stack stored items up to 17 boxes high. The algorithm that runs the operations places rarely ordered items at the bottom and frequently accessed items on the top. This minimises the amount of time it takes to complete most of the company's online orders.

- **Smart roads** – forecasters predict that in the future the use of smart roads with embedded sensors will become commonplace. In the United States and Europe a number of trials have been undertaken with mixed results. Typically, the smart road is made of durable solar panels that need to withstand loads of up to 250,000 US pounds, and are generally made of transparent tempered glass tiles. Underneath these tiles are photovoltaic cells that collect solar energy, which can then be used to power nearby buildings, streetlamps and other public infrastructure. The tiles are also equipped with LED lights, the uses of which include, but are not limited to, marking traffic lanes for controlling or redirecting traffic and conveying messages to motorists such as warnings about upcoming hazards. As the road panels can be heated, they will also be able to stay ice-free during the winter months. It's worthwhile saying that they are far more costly than conventional road surfaces.

7.1.6 Retail

The retail sector is one of the largest private sector employers worldwide and is made up of businesses that sell goods through stores and on the internet to the public. It is a diverse industry sector, and includes everything from department stores to restaurants to pet food shops, selling consumer goods or services to customers through multiple channels of distribution.

The retail industry is using AI in many ways, including:

- **Sales and customer relationship management technologies** – to target and retarget customers with upselling, cross-selling and similar buyer recommendations.
- **Delivery drones** – the much discussed but long-awaited arrival of the AI-powered delivery drone is still a way off. Several organisations, including Amazon and DHL, have trialled the technology but face a number of hurdles to the acceptance of this form of delivery.
- **In-store chatbots** – chatbots can increase your customer experience, help you find things around a store and, while taking you to the items you are looking for, can also monitor inventory/stock levels on shelves.
- **Cashier/cashless stores** – Amazon developed its revolutionary Amazon Go store to allow customers to simply walk into the store, take what they want from the shelves and walk out without going through a cashier. This is possible though sensors and cameras placed throughout the store that track what customers purchase, and their Amazon account is charged when they leave. This is an example of AI helping to create a quick and seamless shopping experience so customers aren't stuck waiting in a queue to pay.
- **Visual search** – this AI technology allows people to take a photograph of a product that they can then search the store's inventory for.
- **Stock control** – stock checking, control and replenishment ranges from the use of robots checking and replenishing shelves to the management or coordination of a warehouse or transport network. The store then uses AI to analyse store receipts and returns to evaluate purchases at each store. The algorithm helps the store to know what items to promote and stock more of in certain locations.
- **Smart stores/fitting rooms** – shops are starting to use AI to run touchscreen smart mirrors, which allow customers to browse through clothing items and gain inspiration. Shoppers can then try them on in an interactive fitting room with custom lighting options. The fitting room mirrors use radio frequency identification (RFID) technology to automatically know what customers are trying on and tell them what other colours and sizes are available.
- **Personalised advisors** – are the use of AI to provide consumers with help regarding a purchase, be it advice about what coat to wear for a particular occasion or to match skin tones for makeup or skin sensitivity.

7.1.7 Media and entertainment

AI offers huge promise for both media and entertainment businesses. AI is currently touching everything from content creation to the consumer experience. AI in entertainment will be used for marketing aspects such as design, film promotion and advertising – smart algorithms will be able to come up with the best marketing and personalised advertising solutions.

AI is in use in the following areas of media and entertainment:

- **Branded content production** – AI is making movie production a lot easier as studios and production houses can 'train' AI using various multimedia content, including text, images and videos, which will then be used to automatically generate different design concepts. This way, AI can easily and instantly generate movie posters and promotional videos. At some point, it's going to be able to create a fully fledged trailer on its own – but not quite yet.
- **Improving the workflow in entertainment production** – there are examples of AI in use at nearly every stage of pre-production and post-production as well as distribution. We can now use AI to improve the accuracy of human characters in computer-generated imagery and digital special effects, or to insert and remove objects from a scene.
- **Content moderation** – in both movies and social media AI can better moderate content across vast data sets and repositories of written, verbal and visual content. Using AI, people can completely delegate these tasks to the software. It can then automatically perform comprehensive scans of the images and videos and identify explicit content– radical, sexual, racial or age limited – removing the unsuitable material. On top of that, AI can even process spoken language and trace offensive speech that can be easily censored afterwards.
- **Computer games – multiplayer human and machine games** – video gaming is another form of entertainment that attracts millions of users from every part of the world; people buy consoles, PCs and smartphones to enjoy their favourite games. One of the most instrumental parts of any single-player game is non-player characters (NPCs). NPCs are the artificial characters that fill the gameplay and react to the actual player's movement and actions. For instance, when they see a player, they start shooting at him/her, while when a player starts shooting, the NPCs hide behind walls and obstacles. Current NPCs are, in some ways, 'AI-ed', so they analyse the player's actions and improve their response over time. As this technology improves, games will feature NPCs even while in the multiplayer mode, which is something that current games do not do.
- **Interactive media and news** – AI enables the ability to choose your own adventure or ending. An example is the standalone interactive film from the TV series *Black Mirror*, a single episode called 'Bandersnatch', first aired in 2018 and representing the latest mainstream offering in interactive storytelling. Similar AI technologies have been employed with news, allowing consumers to decide what type and depth of news they want to consume.
- **Social media** – several AI-powered tools already exist to deliver insights from your brand's social media profiles and audience. This often involves using the power of AI to analyse social media posts at scale, understand what's being said in them and then extract insights based on that information.
- **Immersive audience experience** – using AI to allow someone to take on a role within a VR or AR environment.
- **Personalised programming/content selection** – AI is being used by mainstream media to transform content. Examples include the use of AI to provide subtitles, audio and text description and tranlsation. This makes media content more transferable to other markets.

- **Content delivery** – it is commonplace for websites to have suggestions for products, and this technology is now becoming mainstream in online streaming services. Emails and adverts make suggestions for what types of films and soaps you might like to watch next.
- **Fully responsive adbots** – these will be available before too long, to allow interaction with advertising on a personal basis based on AI-marketing knowledge of individual preferences and current circumstances. An example could be interacting with an online or physical display advert that can identify you looking at it and advise you to remember to buy your partner's birthday present next week based on their likes and dislikes, and even suggest where it can be purchased in person or online.
- **Computer-generated imagery (CGI)** – AI can be used to enhance movies by the colouring of black and white images, making them more relevant to today's generation who have never experienced black and white films.
- **Music production** – there are AI programs that can write their own symphonies, assist musicians in lyric-writing and tailor music recommendations to users. This list is expanding as we move forward.
- **Personalised movie content** – AI enables advertising in movies based on your individual preferences.
- **Content production** – for content production that covers the 'create' and 'produce' categories of the content chain, projects are underway to enhance creativity and to produce more content with reduced resources.
- **Synthetic personalities** – powered by AI, these will change the way we learn about new products and how to use them, with the creation of avatars and interactive personas that do not exist in nature.
- **Lip-reading** – Google's DeepMind and the University of Oxford collaborated on a project to apply deep learning to a huge data set of BBC programmes to create a lip-reading system that surpasses professional capabilities. The AI system was trained using some 5,000 hours from six different TV programmes, including *Newsnight*, *BBC Breakfast* and *Question Time*. In total, the videos contained 118,000 sentences.[69] The system has potential use for entertainment or even spying.
- **Deep fake** – deep fake offers the tantalising potential that AI and NNs can generate convincing video without a script or even a storyboard. However, it has been used in a negative way to create events that never occurred, for example the creation of a video showing US President Donald Trump saying something that was said by former President Barack Obama, effectively putting other people's words into a politician's mouth and creating fake news.

7.1.8 Transportation

A recent report on the use of AI in the transport sector stated that AI has the potential to make traffic more efficient, ease traffic congestion, free drivers' time, make parking easier and encourage car- and ride-sharing.[70] As AI helps to keep road traffic flowing, it can also reduce fuel consumption caused by vehicles idling when stationary and improve air quality and urban planning.

The various uses of AI in the transport sector include:

- **Self-driving vehicles/autonomous vehicles** – this is the area that people are most aware of, with several high-profile developments over the last 20 years. Most car manufacturers now have autonomous vehicle programmes in place – AI is used to enable the cars to navigate through traffic and handle complex situations. With combined AI software and other IoT sensors, such as cameras and Lidar, it becomes easier to ensure safer driving. Although many of the technological components exist for an artificially intelligent car today, because of various regulations round the world true autonomy is still many years away.

  There are five levels of vehicle automation:

    - **Level 0 – no automation**. This describes your everyday car.
    - **Level 1 – driver assistance**. Here we can find adaptive cruise control and lane keep assist to help with driving fatigue.
    - **Level 2 – partial automation**. Here automation can assist in controlling speed and steering, although the driver must have hands on the wheel and be ready to take control at any given moment.
    - **Level 3 – conditional automation**. Here hands can be off the wheel, but drivers are still required behind the wheel. Vehicles are capable of driving themselves, but only under ideal conditions and with limitations, such as limited-access divided highways at a certain speed.
    - **Level 4 – high automation**. Here autonomous vehicles can drive themselves without human interactions, but will be restricted to known use cases.
    - **Level 5 – full automation**. Here we have arrived at true driverless cars. Level 5 capable vehicles should be able to monitor and manoeuvre through all road conditions and require no human interventions whatsoever, eliminating the need for a steering wheel and pedals.

- **Public transport** – AI in transportation and infrastructure can collect traffic data to reduce congestion and improve the scheduling of public transport. Transport is affected by traffic flow, so AI will allow streamlined traffic patterns, smarter traffic light algorithms and real-time tracking, which can better control variances between higher and lower traffic patterns.

- **Traffic management** – AI is being applied to help identify traffic patterns, manage traffic lights and improve road safety. What is becoming known as 'Traffic AI' is being developed to allow systems to collect and analyse traffic data, generate solutions and apply them to the traffic infrastructure, creating 'smart roads'. The three main areas of development of AI systems are in managing traffic light systems, creating smart traffic lights and understanding the traffic patterns using predictive analytics, which can then help to prevent or alleviate road congestion before it occurs.

- **Improvements in public safety** – worldwide, approximately 1.35 million people die in road crashes each year; on average 3,700 people lose their lives every day on the roads and an additional 20–50 million suffer non-fatal injuries.[71] So, another area of Traffic AI development is in improving safety through better detection of road

traffic disturbances in real time and integration with smart AI-enabled vehicles to allow the detection of pedestrians about to cross a road and prevent accidents.
- **Delay predictions** – in the future, reliable AI may be capable of autonomously controlling traffic flow across a whole city by building AI into road infrastructure and autonomous vehicles, creating a truly advanced traffic management system.

7.1.9 Engineering

AI and ML have made inroads into engineering for many years, both in hardware, with robots used in automotive production, and software, in simulation and modelling. Using AI has allowed engineers to focus on the human interface, leaving AI to focus on the number crunching. Many people, including engineers, fear the impact that AI will have, but it will free up humans to do higher-level tasks, as well as take over jobs that require the unique skills of humans, some of which don't even exist yet. Engineering is at the forefront of this brave new world.

Some key areas where AI is used in engineering:

- **Robotics** – this is the first area of engineering people think of when looking at the impact AI has had and is having. Autonomous robots can work in extreme environments where humans cannot operate efficiently or safely, for example in sterile, vacuum or underwater environments or in chemical, radiological, biological or physically hazardous conditions. Also, robots can operate 24 hours a day and 365 days per year and don't yet require holiday pay.
- **Manufacturing** – AI has made huge inroads into manufacturing in several areas, such as creating smart supply chains to allow the stock reduction of parts having to be held and thus creating lower stock costs. AI is also used in the detection of defects in goods and products, and in the reduction of downtime in manufacturing equipment. This has led to the development of smart factories and production lines.
- **Design** – with AI-powered generative design software algorithms, AI has the potential to create innovative designs by generating many design permutations and alternatives based on an initial set of requirements and constraints. ML will then decide on the most appropriate design solution to adopt. It has been used in the design of many contemporary buildings.
- **3D printing** – AI has enabled the development of 3D printing of parts, and more recently buildings – engineers are even planning how buildings could be 3D printed and assembled on Mars by robots.
- **Programmable robot swarms** – these are simply multiple robots working towards common goals using collective behaviours. By observing how artificial or natural systems (flocks of birds, shoals of fish, ants) work, engineers say it is now possible for AI machines to learn how to collaborate and cooperate in swarms.[72] Potential uses for swarm robotics include deployment for search and rescue tasks, and environments that require miniaturisation (nanobotics, microbotics) for distributed sensing tasks in micromachinery or even inside the human body.

- **Internet of Things** – IoT has become an engineer's playground. IoT is devices interacting using the internet, which allows flow of data between devices. AI can help to make sense of these data, making the devices learn from their data and experience. Coupling AI and IoT together (the Artificial Intelligence of Things – AIoT) is purported to be the basis for a whole new level of engineered solutions. In simple terms: your doorbell will be able to engage with a courier delivery driver without you having to be involved; your intelligent home thermostat will understand your patterns of activity, behaviours and preferences by linking to your diary and the location of family members and will modify your home heating, ventilation and air-conditioning systems accordingly. In the future this level of intelligence will be engineered into all new smart homes rather than bolted on.

- **Virtual and augmented reality** – VR/AR systems (blending physical and digital environments) have rapidly evolved over recent years to the point where AI models are replacing some of the more traditional computer vision approaches that underpin AR experiences. AI is now doing many of the things required to build immersive AR experiences. AI uses DNNs that can estimate depth and segment images for realistic occlusion, detect vertical and horizontal planes and even infer 3D positions of objects in real time.[73] This allows fully immersive environments to be viewed without actually having to build them.

- **Simulation modelling** – many organisations are already starting to apply AI to their digital supply chains, smart factories and other industrial processes, and include AI in their simulation models by creating digital twins. The AI can be directly embedded in the simulation model to facilitate testing and forecasting, allowing organisations to tailor and analyse their systems in the digital world before applying it to real-time systems. Agent-based systems often require a lot of parameters, so it makes sense to employ the power of AI to manage the number of permutations and reduce the overall processing times of producing simulations. Applying AI to optimisation and calibration is another key opportunity in simulation modelling. Being able to simulate the movement of people, or even smoke, around a building before it is built is a huge benefit to architects and engineers. Simulation can also be used to train AI systems.

- **Control automation** – control engineering is a means of managing and measuring the performance of process systems in industrial areas ranging from construction companies and manufacturing plants to power generation plants and nuclear reactor sites. One application of AI in control automation systems is using the data from various machines, sensors and robots to help improve safety standards by learning what was done in the past and then applying it to future situations, thus preventing major incidents.

- **Predictive maintenance/product nervous system** – rather than waiting for a component or product to fail, or undertaking unnecessary maintenance, large data sets can be processed using AI to identify patterns and insights with minimal or no human intervention. The AI will learn from the component or product or IoT and predict when something is about to break or is in danger of breaking. For example, an AI system accessing various sensors on an aeroplane's engine throughout its life could predict when something was due to occur by learning how the system operates and preventing what could be a catastrophic incident. The technology is already available to make these calculations and help to troubleshoot if there is a problem. Another example uses ML: you can prepare a model of a factory, identify normal and abnormal states and then, with the use of real-time sensor data, predict degradation in a manufacturing process or production line and take appropriate action to avoid inefficient operation and potential downtime.

- **AI hardware** – this is optimised hardware used for ML mathematical operations, for example convolution. The hardware includes:
    - GPUs – graphics processing units;
    - OPUs – optical processing units;
    - CPUs – central processing units;
    - TPUs – tensor processing units (hardware for linear algebra operations in AI).

7.1.10 Agriculture

With ever increasing concerns around worldwide food security with a growing population and climate change challenges, AI technology use in agriculture is helping farmers to improve their efficiency, to reduce hostile environmental changes and poor plant nutrition and to aid detection of diseases and pests on farms.

AI uses in agriculture include:

- **Weather forecasts** – AI is now producing more accurate weather forecasts with greater levels of granularity, allowing farmers to more accurately plan sowing and harvesting and improve use of scarce water resources.
- **Crop and produce planning** – AI systems are helping to improve harvest quality and analyse farm data by assisting farmers to optimise planning and generate better yields through determining crop choices and resource utilisation.
- **Analysing farm data/precision agriculture** – with access to thousands of data points, farmers are increasingly using precision agriculture to improve accuracy in seasonal crop yield forecasting by using probabilistic models.
- **Soil quality and water tracking** – analysing soil quality can often be an expensive and lengthy process. AI has come to the rescue. Researchers at IBM in Brazil created a prototype AI-powered smartphone app and the 'AgroPad', which is a paper device about the size of a business card.[74] It contains a microfluidics chip that can perform chemical analysis of a water or soil sample in less than 10 seconds. The farmer simply puts his sample on one side of the card, and on the other side a set of circles provides colorimetric test results that, when scanned by the phone app, advises the farmer of the soil quality regarding a variety of nutrients.
- **Harvest improvement** – several agri-tech companies are now quickly developing agricultural yield boosting algorithms that can show farmers what resources and techniques are going to be best for a crop.
- **Using robotics** – AI-enabled cameras attached to drones can capture images of an entire farm and analyse the images in near real time and at different wavelengths to identify problem areas and potential improvements. AI-enabled agriculture bots also help farmers to find more efficient ways to protect their crops from weeds, either by picking out the weeds or with more targeted use of herbicides.

- **Tackling labour shortages** – agri-tech companies are working on field robots to sow, tend and harvest various crops to address the worldwide agricultural labour market shortages, especially around seasonal crops such as strawberry picking.
- **Pest and disease control** – with an image recognition approach, AI identifies possible crop defects and pests through images captured by a camera.

7.1.11 Military

The use of AI in a military context is highly contentious, but the genie is truly out of the bottle so we should be aware of its current and potential use. The precursors to AI, such as operations research, have been employed in various military contexts since the Second World War.

Various reports on AI have been produced in several countries: for example, in France the Villani report of 2018 stated that the use of AI will be a necessity in the future to ensure security missions, to maintain power over potential opponents and to maintain France's position relative to its allies;[75] similarly, the United States produced a report in 2019 on AI and national security.[76] Most countries have produced similar reports in the last few years and in most reports the use of AI in military contexts has raised ethical issues, which are still to be addressed.

Current and potential future military uses of AI include:

- **Autonomous operations** – AI technology could, for example, facilitate autonomous operations, lead to more-informed military decision making and increase the speed and scale of military action.
- **Intelligence, surveillance and reconnaissance** – the amount of data produced by the various branches of the military have increased exponentially, and it has become impossible for human analysts to sift through the sheer volume of data from the various data feeds. AI has come to the rescue with AI algorithms doing the heavy lifting by performing pre-processing and automation tasks, allowing analysts to focus on analysis rather than data processing.
- **Military drones/unmanned aerial vehicles** – the military has used drones since the early 2000s for a number of purposes. They range in size from hand launched drones to pilotless fighter aircraft. One use of drones in intelligence, surveillance and reconnaissance is where various AI technologies, such as machine vision, are employed, allowing drones to fly on their own. The AI behind the drone therefore needs to be trained using a supervised learning process to identify various objects such as buildings and enemy targets.
- **Battlefield robots** – with the development of AI-powered military systems (humanoid robots and tanks so far), the US military is now on the verge of deploying machines capable of going on the offensive on the battlefield, picking out targets and taking lethal action without direct human input.[77] At present, military officials have not given machines full control and there are no confirmed plans to do so. Does this apply to all countries with this capability? We don't know.

- **Lethal autonomous weapon systems** – lethal, and largely autonomous, weaponry is not entirely new. A handful of these systems have been deployed for decades, though only in limited, defensive roles such as shooting down inbound missiles hurtling towards a ship and autonomous ships hunting for submarine targets. The US Air Force is currently working on an autonomous F-16 fighter jet as part of its, provocatively named, Skyborg programme.[78]
- **Cyberspace operations** – protecting a country's national networked data and communication systems has become a key element of national defence capabilities, and AI is playing a large part in detecting and blocking attacks against key elements of national information technology (IT) infrastructure. New AI applications are emerging on intent-based networking security, on AI platforms for cyberdefence and on immune computer systems that can self-adapt to changing cyberattacks. This area of defence has become a game of cat and mouse.
- **Logistics** – military logistics is a huge area where AI can make a massive difference by managing the millions of assets that the military need to deploy. The task means large quantities of data to sift through in order to make decisions regarding supply, transport, communications and so on. Using AI could certainly help speed up the process and make it more agile.
- **Command and control** – the US *Department of Defence Dictionary of Military and Associated Terms* defines command and control as: 'The exercise of authority and direction by a properly designated commander over assigned and attached forces in the accomplishment of the mission.'[79] The military will make use of AI tools, which will be required for analysing, managing and making use of different pieces of information, to provide easily achievable advantages. They will require detailed planning, so they will need AI tools for working with tactical databases (terrain, logistics, etc.) and AI decision support tools to make it possible for the commander to evaluate different courses of action at different abstraction levels.

7.1.12 Sales and marketing

With AI it is possible to make all sales and marketing processes much faster by utilising tools such as predictive analytics. The positive influence of data science on sales and marketing has grown enormously over the last few years. AI is now helping marketers to collect data, identify new customer segments and identify and use buying intent data to target potential customers. Overall, AI is allowing us to create a more unified sales/marketing and analytics system.

AI can now scale customer personalisation with precision in ways that did not exist until recently, by categorising customers into distinct personas and understanding exactly what motivates each persona to make a purchase. With this information in mind, marketers can focus on the specific needs of their audience and help to establish a long-lasting relationship between the customer and a particular brand.

Specific areas where AI is used in sales and marketing are:

- **Predicting trends** – with AI technology, marketers can spot microtrends and even predict trends. They can then make strategic decisions about where they allocate their budgets and who they target. As a result, brands can reduce digital advertising costs and ensure that their marketing budget delivers the best possible results.

- **Coupling AI with intent data** – intent data are digital signals buyers emit as they're moving through the purchase journey; data pinpoint where buyers are in their path-to-purchase and can be a powerful predictor of who to target and when. AI is key to accurately analysing these signals at scale. This is a revolutionary approach, helping marketers to identify, engage and nurture the right buyers at the right time with the right message. For example, the AI system can identify customers that leave items in their shopping basket and then identify what additional offer best works to get the customer to log back in and make the purchase.

- **Building pipelines** – AI tools can now help salespeople fill their sales pipeline faster than ever by identifying new leads within existing databases and new leads that look like current leads, even to the point of providing the contact details for people in specific markets.

- **Closing leads** – by using AI predictive analytics it is possible to target and rank individual leads based on their likelihood to close, using what the AI tools have learned from previous successful sales. This helps to prioritise workload on only the most promising leads, which increases close rates and boosts sales figures.

- **Auto sales chatbots/avatars** – telemarketing roles are gradually being replaced with various AI-powered bots that are now available for providing sales advice lines and answering frequently asked questions, and even making personalised outbound marketing calls to engage with leads human sales staff do not have time to contact personally, potentially 'warming up' the lead for a human follow up call.

- **Automating routine sales tasks** – AI tools can create customised emails that are specific to a prospect and help sales staff to arrange customer meetings. AI systems can even suggest responses during email interactions or during live conversations.

- **Sales data analytics** – AI tools can provide data analysis by reviewing data from past sales events and identifying patterns in product performance and purchasing patterns. The AI systems analyse data, make assumptions, learn and provide predictions about customers, opportunities, demand and supply, at a scale and depth of detail previously impossible for human analysts.

- **Product pricing** – AI will change the way that we do product and service pricing. Analysis of different data sources will help to make decision makers more effective and efficient in setting prices, creating an effective pricing strategy, calculating sales promotions and the return on investment, and making customer segmentation more accurate. AI software is already available to offer automated and dynamic pricing, it can collect competitive pricing and offer customers different personalised prices based on external factors (availability, geographical location, brand loyalty, etc.) and their individual buying habits. The same AI applications can also show price fluctuations over a period of time.

7.2 SUMMARY

Many other industry sectors, including construction, mining and petrochemical, are investigating the use of AI to become more competitive, efficient and cost-effective. Although we have not covered them in the use cases detailed in this chapter, it does not necessarily mean that they are not actively pursuing the use of AI. At present it's widely reported that 27 per cent of organisations are in the discovery/experimentation phase of AI,[80] inferring that many industries are yet to start adoption of AI 'best practice' and those who have woken up to the opportunities that AI brings have still a long way to go.

8 AI CASE STUDIES

In this chapter we look at actual real-world case studies of AI in action; real-world problems where learning from experience of humans and machines is required. We have to balance the limitation of AI theory, the limitations of AI technology and what we can actually learn from experience practically. We are not going to go deep into the mathematics, however, because we would drift towards ML too far and lose the big picture.

The first case study is an ML problem, the second is an AI problem with an important role for ML and the final case study looks at the hard work of engineering AI technology in a company in the early, embryonic, stages of growing.

8.1 PREDICTING THE PERFORMANCE OF A WAREHOUSE

In this section, we start with an example of operational research and engineering, two of the underpinning academic disciplines that ML comes from. This first study examines the loading of a warehouse storing products that need to cool over a long period of time. It is based on a real example, and we will assiduously stick to the process of what was done. Our aim is to introduce you to what is involved, practically, in developing a project where 'learning from experience' is key and AI or ML can take away the heavy lifting – in this case, that of monotonous calculations.

8.1.1 Background to the problem

On a large industrial site space is precious, and building a large store able to allow products to cool is costly. Storage costs must be kept to a minimum, which drives the incentive to produce a building design that allows the products to cool naturally.

In this case study, heat from the contents was used to drive buoyant, warm air out and draw cool air in, which imposed strict limits on the total amount of heat that could be held in the store at any time. Replacing the mechanical cooling equipment was safer, better for the environment, quicker to build and easier to operate, which made the design of the naturally cooling stores a sensible option from a sustainability point of view. However, the design team were unsure of the inventory, so they needed to learn how to model these aspects of the warehouse.

As the storage vault could only hold a set maximum heat, the next question was how to represent the cooling of the vault's contents in the model. A busy industrial site is a complex environment with lots of activity; so, should they model the movement of the products around the site and assess how they have cooled before reaching the store? Should they model the local weather and take advantage of the colder winter months?

If we are not careful, conundrums can run away with themselves and quickly turn a model into a complex system – the number of options becomes overwhelming! Is this an AI project or a ML project?

8.1.2 Is it an AI project or an ML project?

In this book, we have assumed that AI is based on the agent schematic (learning, sensors, actuators and an environment) and that ML is learning from data. In this case study there is certainly an environment that will contain lots of sensors and actuators, and we could possibly learn from these. We can make some assumptions to simplify our problem and enlist experts to give us the 'heuristics', or 'rules of thumb', that will make our learning from experience much more effective. It certainly could be an AI project, but it is a ML project – learning from data!

8.1.3 There are simply too many things to consider

At the start of a project like this, there is a temptation to build a solution that is too big and too complex; that is called a combinatorial explosion. In addition, because a computer can do lots of calculations quickly, there is also a temptation to use brute force and model every permutation. This is unworkable, in reality, because we have limited computational hardware and will somehow have to analyse all the results. For example, it is unfeasible to model every possible move in chess using brute force.

To begin work on a project, we need to understand what constitutes 'fit for purpose', and so need to consult the domain experts. In doing so, we can also ask what rules of thumb we could draw on to give us a starting point. In this case study, the domain experts kept the project's goals realistic and, after significant discussions, the following concepts were developed:

- The key parameter was the total heat load in the vault.
- The heat load of products going into the store needed to be known.
- The rate that each type of product arrived at the store needed to be known.
- The rate that heat generation decayed needed to be known.

To avoid the combinatorial explosion, the human 'learning from experience' needed to set up a project that could be solved by machines. At this point, the team had a pretty good idea of the scope of work – the domain expert had given their best starting point and a sensible way to draw back from a combinatorial explosion.

It was clear this was a ML problem rather than AI because the environment was not being manipulated with sensors and actuators – ML was being used to design the building by:

- understanding the rate at which each type of product arrived at the store;
- modelling the decay of heat generation of the products in the store.

AI CASE STUDIES

8.1.4 The scientific method

The scientific method, described in Chapter 1, can be used to understand the two ML parts of this project.

Note that the ML was going to learn from data about the heat generation characteristics of each product type; it was also going to be rigorous, fit for purpose and reproducible. The domain expert could quality assure it. The team avoided an intractable problem, gave themselves a good chance of success and built domain expert credibility. The plan also placed them on a good foundation for justifying the work with stakeholders.

8.1.5 Is it technically robust and trustworthy?

At the early stage of the project the team could not be absolutely sure whether the product would be technically robust and trustworthy without analysis and the results. What they could say was that the adoption of the domain expert's heuristics, a good understanding of what 'fit for purpose' meant and using the scientific method pointed them in the right direction.

The two ML project activities now needed another domain expert: a product specialist who could define and provide models to predict the heat generation rate of each product and, of course, when the products were going to arrive.

8.1.6 The machine learning solution

> If AI or ML projects go wrong (e.g. they are too costly, too complicated, too slow), they need to be stopped quickly and reassessed. Human and machine learning must be done in parallel, and the human learning can often be overlooked.

The product specialist needed to use ML to characterise the products into six different types. In doing so, the expert also determined that the products would arrive between fixed start and end dates. The rate at which the products arrived was also uniform, which was ideal as it made the prediction of what products would be in the store at any point in time almost trivial. At this point, the ML could be considered to have done its job, but there was another task to undertake: how to model the cooling of the product.

To make this task simpler, the team made the assumption that, provided the store could safely remove heat if it was not overloaded, they only needed to know how much cooling was required at any point in time. The products generated heat because they were undergoing exothermic reaction. The heat generation was not trivial, but the domain expert was able to provide the heat generation decay curves as data points and the ML modelled them, which is best described as numerical analysis. Simple cubic splines were used to give a piece-wise continuous representation of the decay of heat generation for each product type.

A simplified model of the heat load within the store was now in place. It used the scientific method to justify its results. By working with the domain experts, the team managed to simplify the ML and avoid a runaway problem, while making sure they did not fall into the trap of using the latest buzzword ML technique. They arrived at this point using human learning from experience as well as enlisting the use of computers and ML to avoid repetitive and error prone calculations. The higher value work within this project was undertaken by the humans, and the machines did what they do best – the monotonous accurate calculations.

8.1.7 The output of the project

It took about six weeks for the project to be completed. The output was a report, not particularly long and therefore easily digestible by the stakeholders, reviewers and engineers. The report was written as the project was running, rather than at the end, and it told a story constructed using the scientific method and backed up by the results. Learning from experience was used wherever possible by both humans and machines; heuristics and assumptions allowed the problem to be simplified so that the hardware was sufficient to actually obtain ML results and the team avoided a combinatorial explosion of possible outcomes.

Of the six weeks, it took about four weeks to prepare the ML for the task; this included identifying domain experts and obtaining their input and results. The final two weeks used the ML to model the store's performance and complete a few learning iterations to optimise the store's design.

The ML clearly demonstrated with the first few sets of results that the proposed store would overheat within a short period of time and that it would take 6–12 months for it to cool to a point where further filling was an option. Engineers quickly realised that they could modify the layout of the store and provide two vaults, each with their own natural cooling. The final report showed how a single building store with two vaults could be used to accept all of the products it needed to. One vault would be filled to its heat loading limit and then the second vault would begin to be filled. While the second vault was filling, the first vault was cooling and the heat load reducing. When the second vault reached its limit, the filling reverted back to the first vault and so on.

The report included one graph within it to visualise the results. An example of the type of graph is shown schematically in Figure 8.1 (these are not the actual results from this project). We can see that Vault One hits its limit and then cools while Vault Two is loaded. Vault Two hits its limit and then Vault One is loaded again and so on. There are about four swaps from loading one vault to the other.

The review panel and engineering team were very happy with the report; it saved them having to construct two stores, and minimised the modifications to the design of the existing stores.

8.1.8 Where's the learning from experience?

On this project the team learned something from domain experts and then constructed a ML project that was simple enough to run with existing hardware to solve it. Human

Figure 8.1 Store's heat load: Vault One and Vault Two

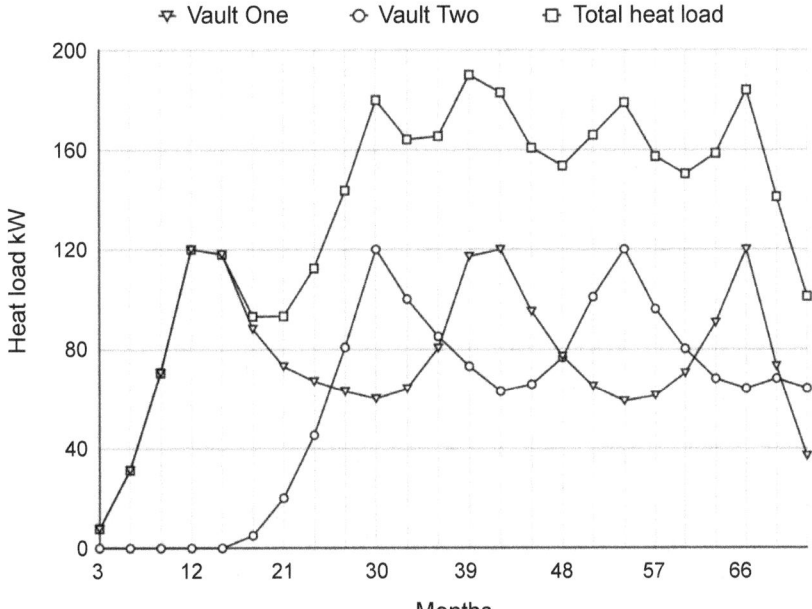

learning started at the beginning of the project and didn't stop until the report had been signed off and accepted, and the human learning both before and during the project was used to avoid an overly complex ML project. The ML was used to learn from data about the products being placed in the store. It gave insight so the team could predict the heat loading in the store and how that store would cool over time. The ML took away the monotonous and tedious calculations, leaving the humans to concentrate on the higher value aspects of the project. It also increased the accuracy of the calculations.

Once the ML had given a better understanding of the product data, engineers were able to make slight modifications to the storage design and reduce the total number of buildings needed, thus reducing their footprint and saving time and cost.

At the core of this ML project were the domain experts who gave the heuristics and assumptions necessary to simplify the problem. The engineers could focus on two fundamental ML problems:

1. Understanding the rate at which each type of product arrived at the store.
2. Modelling the decay of heat generation of the products in the store.

It would have been easy to build a more sophisticated ML project, but that would have been likely to hit a combinatorial explosion requiring more computing power than was available.

> **THE REAL WORLD ISN'T A SIMULATION**
>
> What would happen if we designed everything digitally? Would you trust something that is designed entirely by digital modelling? In this box we are not going to concentrate on a single example, but look at a concept that has been around for about 20 years – that of the digital twin.
>
> The digital twin is an example of where AI can be used to learn from a digital simulation, and maybe sensors; there are many examples where this concept has been used successfully.[81] We also need to note that just because we have simulated something, it doesn't mean it will perform the same way in reality.
>
> The digital twin has various definitions, and is not something to start without considering carefully if it is a good representation of the real world. In the next case study, we show how even a simple model of the Earth's weather turns out to be incredibly complicated and requires large supercomputers to undertake.
>
> AI can also draw on the system of systems approach to modelling complex systems – lots of subsystems that, when combined, develop ever more complex behaviour. This has the advantage of being an academic subject linked closely with the fundamental disciplines that underpin AI and ML. Examples of these approaches are agent-based modelling and life cycle analysis used in sustainability. With a system of systems approach, you can break the larger problem down into manageable chunks. Indeed, we might find out that one system requires so much computing power that the larger digital twin might not be feasible, or indeed realistic! In this case we need to build a model of the subsystem that can be incorporated into the system of systems approach.
>
> The digital twin concept built using the system of systems approach might be useful in ensuring that our AI is human-centric, technically robust and trustworthy.

8.2 THE WEATHER – CONCENTRATION ON DATA PREPARATION

In 2020 the European Centre for Medium Range Weather Forecasts (ECMWF) had the computing power and human resources (one of Europe's largest supercomputers and a few hundred dedicated and talented scientists, engineers and mathematicians) to produce weather forecasts ranging from 15 to 365 days ahead.[82] The UK Met Office has a similar make up.[83] The knowledge base required to simulate the weather and climate requires the physical processes to be represented mathematically and solved numerically plus the knowledge of how to build the hardware to do the heavy lifting!

Both the ECMWF and the UK Met Office have to model a complex system with limited computational resources; current models of the weather have a resolution of 10km, plus both organisations monitor and obtain data from numerous weather stations and other facilities worldwide. Supercomputers are not powerful enough to process the simulations and data in a timely fashion – the sensors are not coupled together, so data does not arrive in an ordered way. The majority of the data that simulations produce is simply deleted because they cannot be stored, so it is impossible to simulate exactly the fluid movement, heat transfer, electromagnetics, chemistry and so on. To get around this forecasters have to build models and capture data from weather observations. They need to build models that capture the non-linear and random nature of the Earth's atmosphere that can be used in simplified simulations. Not only that, they have to analyse and process more data than could ever be understood by humans in many lifetimes – and within hours, not days. They need machines to help – this is not a nicety, it is an emphatic necessity!

In this case study we are going to look at the modelling of the turbulent fluid mechanics in our atmosphere and, in particular, whether we can simplify the modelling of it. By this we mean: can we build a simple model and use less computing resource? We will always have limited machine resources. The study draws on a paper by Professor George Batchelor in 1969 where a hypothesis about turbulence was tested on a computer.[84] This is an early example of machine learning in science. The hypothesis is still at the core of atmospheric fluid modelling.

8.2.1 Is this a ML or AI project?

Our brief introduction to weather modelling may have given away the answer to this question. Weather forecasting is an AI project because it uses a combination of sensor, simulation and learning from experience. ML also plays a large role in weather forecasting, though, often, lots of weather predictions are combined to build statistical understanding of the forecasts.

> Where are the actuators? Well they are not easy to see, but they are the factories, cars and homes not producing detrimental by-products! If we change our behaviour because the forecast tells us something, we humans are actuators also. Other actuators are the UK Met Office's weather app: it gives us vital information about our weather and climate. In fact, wherever we go, the Met Office weather app usually has access to a local forecast and readings from local weather stations.

8.2.2 Turbulence – Leonardo da Vinci to the Red Spot on Jupiter

In this case study, the domain experts followed the same methodology as Batchelor and his postgraduate research student had used in 1969 and, around 2000, did this work again, this time generating gigabytes of data to test a hypothesis (albeit a different one), just as Batchelor had, but using one of the most powerful computers in the world. The case study is an example of using machines to do research in one of the most challenging areas of research, that of turbulent fluid flow.

Leonardo da Vinci drew sketches of the vortices generated at the bottom of a water outlet some 400 years ago. This is often used as the starting point in studies of turbulence. After 400 years we have moved further forward in our understanding, but we still require a strong and dedicated research effort to do so. Turbulence is complicated because it is random (statistical), enhances mixing, has vorticity and is non-linear. If you want to improve the mixing of cream into a cup of coffee, you stir it up and introduce some turbulence. Turbulence is ideal for chemical engineers stirring a reacting tank, but unnerving for us when we are on a flight.

If we travel on an aircraft we feel turbulence; it's powerful. The physics of turbulence are complex, highly mathematical and studied in every university. To an engineer, it is useful for mixing or maximising the flow in a pipe, but this often comes at a price – it takes energy to do it. What this means is that kinetic energy from the main body of the flow is passed down to become heat at the molecular level. Turbulence is really good at this and we call it the inertial cascade.

An example is the best way to show this. Figure 8.2 shows Niagara Falls, on the border between the United States and Canada. Approximately 675,000 gallons per second, or about 2,547,486kg per second, of water falls 51 metres. At the base is a bowl, and the water hits the bowl at about 71mph or 31.6m/s. By the time it leaves the bowl, the potential energy that was transferred to kinetic energy has been dissipated and is now heat. It is this mechanism that was first sketched by Leonardo da Vinci, and what we see are eddies, large vortices mixing, interacting with each other and being stretched. The stretching causes them to spin quicker and the internal rubbing (or friction) of the water turns kinetic energy into heat. This is a very simplified description of turbulence. It happens very quickly.

Figure 8.2 Niagara Falls – the turbulence here dissipates enough energy to power an area the size of north Wales (reproduced from https://commons.wikimedia.org/wiki/File:Niagara_Falls_Canada_NARA-68145149.jpg under Wikimedia Commons licence)

Figure 8.3 shows an image from NASA of the movement of the oceans. Something we observe in the atmosphere and oceans of planets is that large structures persist; there is something about the atmosphere and oceans that stops the inertial cascade from happening. The famous example is the Red Spot on Jupiter; there's also the Blue Spot on Neptune and the Earth has its own polar vortices as well. The atmosphere of the Earth is like a soap film: it is very thin and promotes the fluid only moving in two directions. This limits, in fact eliminates, the vortex stretching we described in the Niagara Falls example. Also, because the Earth is spinning, and the atmosphere and oceans are stratified, this also promotes movement of fluid in two directions. We call fluid that can only move in two directions two-dimensional fluid flow or two-dimensional turbulence. It allows large-scale structures to form and persist for long periods of time.

In 1969 George Batchelor wrote a paper for *The Physics of Fluids* journal proposing to test a hypothesis about the energy spectrum in two-dimensional turbulence using a computer program.[84] This is a ML problem from more than 50 years ago. He asked a postgraduate student to write computer code to generate the statistical data – ML was being used within the scientific method to do world-class, leading-edge research. These techniques are being used today on the world's fastest supercomputers. The work aims to understand physics that would be impossible to understand experimentally because of physical limitations or costs. It is an amazing example of where ML is doing the heavy lifting while the domain experts, in this case academics, are testing their hypotheses.

Figure 8.3 The movement of the Earth's oceans (NASA's Goddard Space Flight Center, https://svs.gsfc.nasa.gov/10841)

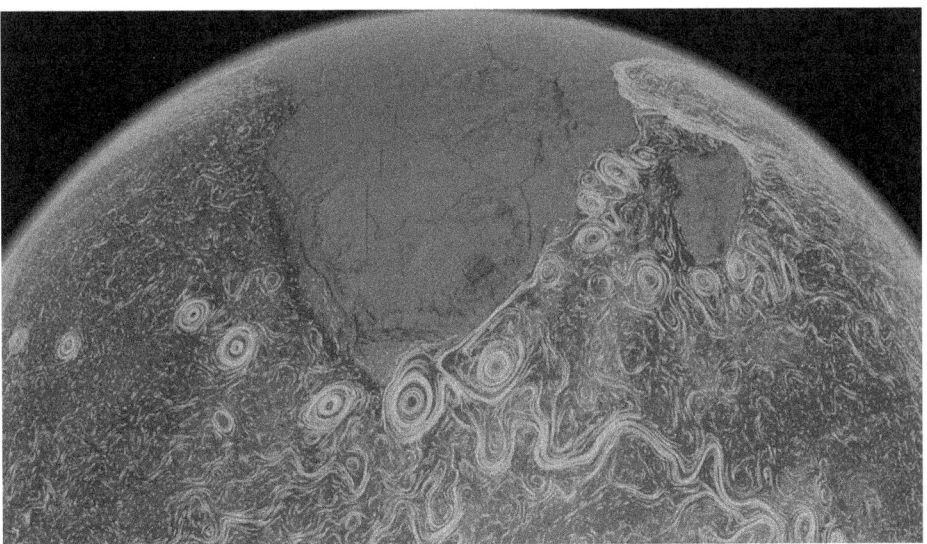

How can we simplify this complex problem and somehow generate data to analyse? The approach uses vector calculus equations, rewritten using linear algebra so we can solve the equations using high-performance computing. The vector calculus equation is

$$\frac{\partial \mathbf{u}}{\partial t} + (\mathbf{u} \cdot \nabla)\mathbf{u} = -\nabla p + \nu \nabla^2 \mathbf{u},$$

where $\mathbf{u}$ is a vector describing the velocity of the fluid, p is a scalar describing the pressure and the Greek letter, ν, is the kinematic viscosity. Letters in bold are vectors and letters in italic typeface are scalars. This the simplest model of fluid flow we can use. In fact, we also had to make the assumption that the fluid is a uniform density and also that it can't be compressed. Vector calculus is short-hand notation, so we can work on the three dimensions without expanding out each term. If we expand the above equation we would have more than 20 terms to deal with. This makes the two-dimensional simplification appealing. With some algebra this equation can be written out in a form that can be solved on a computer – these equations are written such that they can be solved using linear algebra software libraries. Here, we only need to know how the domain experts generate the data we will analyse. In this case, the domain experts, working in 2000, followed the same methodology as Batchelor and his postgraduate research student to generate gigabytes of data.

8.2.3 Too many data to analyse without a machine

It took about 24 months for the domain experts to generate the data; this included building the simulation, running the simulations and writing the software. Visualisation of the data played a pivotal role in the human learning. After all, the hypothesis is developed by the humans for the humans to test.

At its fundamental core, turbulence is a subject that requires a statistical understanding and therefore we need lots of data. The thesis asked the question: How many energy containing eddies do we need in our simulations to represent turbulence well? The supercomputer was used for days at a time, generating hundreds of eddies and letting them merge and interact with each until, after a long time, they had merged into a few large eddies, similar to the ones on Jupiter or the Earth's polar vortices. Figure 8.4 shows a visualisation of one of the simulations. The large data set is visualised using a passive scaler called vorticity to pick out the vortices in the flow. It easier to think of this as watching cream being stirred into coffee. The figure shows the initial, intermediate and final stage of the simulation; and you can see large eddies form, suggesting the energy is accumulating in the large scales.

If energy does not cascade down to the smaller scales, where does it go? Well, it is passed up the scales to form larger eddies. Turbulence theory tries to predict this. It asks questions such as: How will the energy be distributed? How will the velocities in the eddies be correlated? Will the flow conserve angular momentum? Humans built these hypotheses so that we can have simple analytical models of turbulence, heuristics or rules of thumb. The simulations generated huge quantities of data so that we have the scientific information to test our hypothesis. You can see the big picture here: ML is helping us build our hypotheses. The data we need to do this are so vast that we can only use a computer to do it.

Figure 8.4 The visualisation of a large data set from a simulation of turbulence that is found in the Earth's atmosphere or a soap film

At the start of the simulation, lots of eddies interact with each other. We have coloured the eddies with a passive scalar called vorticity, just like stirring cream into coffee.

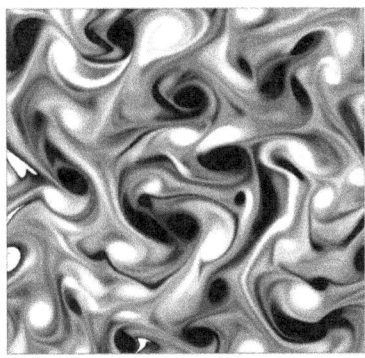

Halfway through the simulation the smaller eddies are merging, and the larger eddies are then sweeping up the background small eddies.

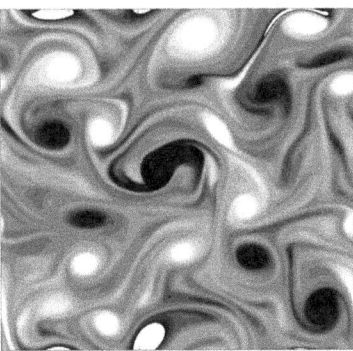

At the end of the simulation we can see that the larger eddies are now well established and more reminiscent of soap film or the Red Spot on Jupiter.

Even with a very powerful supercomputer, we still need to be careful. The simulations are generating terabytes of data. Storage limitations mean that we cannot store it all. In this project only a small fraction of the data could be sampled – well below 0.001 per cent – and, even then, the amount of data were so big that only a machine could analyse them. At each stage of the project we need to follow the scientific method. If we get our hypothesis correct, simple and elegant models could be built that allow humans to understand complex phenomenon such as the weather.

As Stuart Russell and Peter Norvig tell us, AI is a universal subject that can help everyone.[9] Machines can help us to understand long-standing and hard questions about complex systems; and we mustn't overlook the fact that visualisation of results, taking vast amounts of data and putting them into a form that humans can use, is a very important skill for an AI practitioner to have.

8.3 A CAMBRIDGE SPIN-OUT START-UP

In the first two case studies we have seen two problems where we've been able to organise a system into a learning-from-experience problem that computers can help us with. These are, of course, narrow AI problems focused on data analysis. Please don't think that these problems are simple. Far from it. The heuristics and physical understanding of the underlying physics takes dedicated domain experts with decades of experience.

Optalysys are a unique company that spun out of the University of Cambridge Department of Engineering. Their founder had the vision to build a high-tech business in his home town of Wakefield, West Yorkshire, creating high quality jobs and opportunity there. The company are pioneering a new approach to the way we think about representing mathematical operations on machines. Mechanical machines have been overtaken and enhanced by digital computation and now technologies such as optical and quantum computing are enhancing our efforts further. Rather than fitting the mathematical operations to a binary digital computer, Optalysys have gone back to basics and are building machines that do the mathematical operations on dedicated machines.

As Optalysys are an ethical purpose AI company, we should see sustainability at its core. Sustainability encourages us to build a balanced company from the pillars of economic, environmental and social metrics. It almost goes without saying that increasing supercomputing capabilities will undoubtedly be a success in financial terms, but where are the environmental and social metrics?

An example of their economic progress is the growth of the company into a double-digit high-tech employer, and being sited in Wakefield, part of the UK's northern powerhouse, would seem to be in line with some good social metrics. The work they do is the engineering of AI hardware; their team is made up of engineers, scientists, physicists and mathematicians. The company has sponsored two PhDs and funded large sums of money in fundamental research. Sustainability probably seems like a lot of fuss when there are significant financial challenges; we'll talk about funding later. You will see that Optalysys understood from an early stage that the sustainability metrics are a good way to sell their business when raising funds: sustainability is not a quirky subject about feeling good but really challenges us to think about our core values and what we are truly achieving.

The environment metrics fit into the product Optalysys are engineering. There are numerous mathematical operations that their technology can undertake. The one that's useful in AI is the convolution operation. In the convolution of an image, say, with N pixels in each direction, we would need roughly $N^2(1+\log N)$ floating point operations on a typical CPU. Optalysys's technology can do this in roughly one operation. Now this seems impressive, and it is, but the power required to do this operation optically is several orders of magnitude less than a CPU or GPU. So, we now have the environmental metrics: it uses significantly less energy. If, and it's a big if, Optalysys get their technology right, we can reduce the energy consumption of convolution NN training drastically. It's not hard to imagine having today's equivalent of a supercomputer in your smart watch!

8.3.1 What does this technology look like?

Optalysys are combining two techniques, taking advantage of digital computational and analogue computational hardware. In doing so, they tune the hardware to the type of mathematical operation they want to perform. This is a powerful concept. Remember, in order for a supercomputer to solve an equation we have to write it in a way our hardware can do the mathematical operations. It's kind of fitting a round peg into a square hole. The peg fits, but it doesn't quite fill the hole and so spins. When we use computers that use CPUs or GPUs, the processing units rely on discrete sampling of a picture, say, usually on a regular grid or lattice. We then have to undertake lots of calculations at each grid point to obtain quantities like derivatives or integrals. Errors cause us lots of problems and making sure we conserve quantities such as energy or other invariants is easier said than done. If our data don't fit nicely onto a regular grid or lattice, our problems worsen. If our grid is too coarse, then we will not have enough resolution to capture the details we need. If our resolution is too fine, then we will be doing lots of unnecessary calculations. A digital computer may not be the best way to do every mathematical operation that we need. Optical processing can help!

The convolution is a mathematical operation that is an intensive operation on a GPU or CPU. When done using optics or lenses, this operation is performed at the speed of light and at the resolution of the images it is convolving. So, the intensive operation, which is often simplified in a convolution NN that uses GPUs or CPUs, can be done using much less energy, at almost any resolution and at the speed of light. They still have to get the data to and from the processor, but you can see it has some pretty spectacular capability. What is amazing is that the optical processor doesn't need more operations as the resolution increases. If you present the system with a coarse resolution image or an ultra-fine super-high definition image, it uses the same system to do the convolution, whereas a CPU or GPU would need to scale the number of operations with the resolution of the image.

The first system Optalysys built was the mathematical derivative operator; Figure 8.5 is a photograph of that first 'moulding clay and sticky tape' prototype. Figure 8.6 shows the latest prototype using state of the art engineering techniques to manufacture an optical system that can fit inside a typical personal computer case. The next generation is chip sized (see the thumbnail in Figure 8.6); it is only 32mm × 32mm. The company has worked really hard to take the 250kg optical bed and over one metre-long prototype optical system down to a 32mm square wafer you can easily hold in your hand!

There is still a lot of work to do. The progress is impressive and interest from around the world is really encouraging. Their approach is brave; they aren't rebadging algorithms that have been around for years as AI. This could be a revolution in the way we think about engineering our thinking machines. There is always the possibility of the celestial spanner in the works, but the risks appear to be worth it. Only time will tell.

Figure 8.5 Optalysys's first optical mathematical operation was the derivative; the first prototype was built on a 250kg optical bed and was over a metre in length

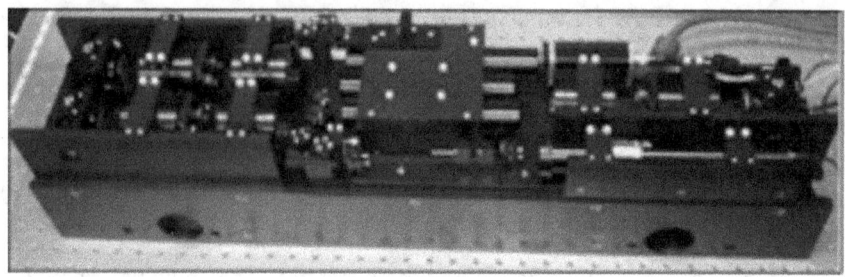

Figure 8.6 Optalysys's first commercial optical AI processor – FT:X Optical co-processor system with PCIe drive electronics. It fits onto a standard PCI card about 100mm wide by 300mm long. The thumbnail, top left, is the next generation of AI processor and measures 32mm × 32mm

8.3.2 Where's the money?

You walk out of a world-class university, hopefully with minimal debt, and have a vision of what your company will look like. Day Two, you realise you need an office, a computer, investors, a bank account, customers, a product. This is not as easy as it sounds, and starts to feel a little daunting. Luckily, Cambridge Correlators (Optalsys were previous called Cambridge Correlators Ltd) had a simple product they could sell at a good price, and their academic background gave them some traction in selling correlators to universities and research organisations. Even large high-tech engineering companies bought them. Investors were already aware of Cambridge Correlators and provided early funding. Grants from local growth hubs gave them access to a nice office and work area. In fact, one grant allowed them to take on their first high-tech graduate.

In the early days it felt like Cambridge Correlators had the solution to an unknown problem. Some possibilities, but the huge inertia of only having early university research at TRL1 or 2 proved to be a significant challenge. The search was on to find 'the problem' so they could build a product to solve it. Conversation after conversation was held with numerous potential customers. The penny eventually dropped that optical technology was actually making working in Fourier or real-space really easy, moving from one space to another at the speed of light. There's more: whole fields can be manipulated to give us access to quantities like derivatives and, not only that, but to combine and manipulate whole fields at the same time in one operation, which is important for AI correlations, convolutions or non-local operators. This glimmer of light (forgive the pun) was enough to allow Cambridge Correlators to transmogrify into Optalysys. They could attack big supercomputing problems such as genetics, modelling the weather, solving equations and, eventually, AI. It took a lot of effort to recognise their own DNA – optical engineering of machines that could effortlessly undertake mathematical operations. Remember, with AI, domain experts need the AI to help them with their heavy lifting. They weren't finding problems; they were making it easy for domain experts to solve their problems. Once they had cracked the DNA of their business, they started writing proposal after proposal. Funding started to come, and the fun of managing cashflow began.

Investors have worked closely with the management team to smooth out the peaks and troughs in cashflow. It is not easy finding investors, especially ones who are patient. The timeline of Optalysys is decades, and this shows how long it has taken to get Optalysys up and running. In the early days a lack of cash could make an opportunity vulnerable. The credibility of Optalysys's use of the scientific method helped in obtaining high integrity grants from worldwide funding bodies. It is what has made this company the unique and inspiring story it is today. The investors have stuck to this principle and not deviated or been fooled by moves towards quick fix whimsical funding opportunities. The scientific method is at the heart of AI technology. They have embraced the machines needed for the next generation of AI systems.

8.4 SUMMARY

These case studies have given us more explicit understanding of the breadth of AI and what is really involved in a project of this nature.

REFERENCES

1. Russell, S.J. (2019) *Human Compatible: Artificial Intelligence and the Problem of Control.* New York: Viking.

2. Heydenreich, L.H. (2018) 'Leonardo da Vinci: Italian artist, engineer, and scientist'. Britannica. https://www.britannica.com/biography/Leonardo-da-Vinci.

3. Watson, R.A. (2019) 'René Descartes: French mathematician and philosopher'. Britannica. https://www.britannica.com/biography/Rene-Descartes.

4. The Editors of Encyclopaedia Britannica (2018) 'Ada Lovelace: British mathematician'. Britannica. https://www.britannica.com/biography/Ada-Lovelace.

5. The Editors of Encyclopaedia Britannica (2018) 'Neil Armstrong: American astronaut'. Britannica. https://www.britannica.com/biography/Neil-Armstrong.

6. Wikipedia Contributors (2016) 'Karen Spärck Jones'. Wikipedia. https://en.wikipedia.org/wiki/Karen_Spärck_Jones.

7. Rogers, K. (2015) 'Tu Youyou: Chinese scientist and phytochemist'. Britannica. https://www.britannica.com/biography/Tu-Youyou.

8. Sternburg, R.J. (2017) 'Human intelligence: psychology'. Britannica. https://www.britannica.com/science/human-intelligence-psychology.

9. Russell, S. and Norvig, P. (2016) *Artificial Intelligence: A Modern Approach*, Global edn. Harlow: Pearson Education Ltd.

10. Kurzweil, R. (2016) *The Singularity is Near: When Humans Transcend Biology.* London: Duckworth Overlook.

11. Dreyfus, H. (1972) *What Computers Can't Do.* New York: MIT Press.

12. McCarthy, J. (1995) Review of *What Computers Still Can't Do* by Hubert Dreyfus. *Artificial Intelligence*, November. http://jmc.stanford.edu/artificial-intelligence/reviews/dreyfus.pdf.

13. Lighthill, Sir J. (1973) 'Artificial intelligence: a general survey'. In: *Artificial Intelligence: A Paper Symposium*, Science Research Council, London.

14. Rumelhart, D.E. and McClelland, J.L. (1986) *Parallel Distributed Processing: Explorations in the Microstructure of Cognition.* Cambridge, MA: MIT Press.

15. Vincent, J. (2019) 'Former Go champion beaten by DeepMind retires after declaring AI invincible'. *The Verge*, 27 November. https://www.theverge.com/2019/11/27/20985260/ai-go-alphago-lee-se-dol-retired-deepmind-defeat.

16 Grinder, J. and Bandler, R. (1976) *The Structure of Magic: II*. Palo Alto, PA: Science & Behavior Books.

17 Parkes, P. (2011) *NLP for Project Managers: Make Things Happen with Neurolinguistic Programming*. London: British Computer Society.
Parkes, P. (2016) *NLP for Business Analysts: Developing Agile Mindset and Behaviours*. Swindon: BCS.

18 Warnock, M. (2006) *An Intelligent Person's Guide to Ethics*. London: Gerald Duckworth & Co Ltd.

19 Team Responsible Robotics and Artificial Intelligence (Unit A.1) (2019) 'Ethics guidelines for trustworthy AI'. European Commission. https://ec.europa.eu/digital-single-market/en/news/ethics-guidelines-trustworthy-ai.

20 Floridi, L. and Cowls, J. (2019) 'A unified framework of five principles for AI in society'. *HDSR*, 1 July. https://hdsr.mitpress.mit.edu/pub/l0jsh9d1/release/6?readingCollection=72befc2a.

21 Russell, S.T., Bakken, R.J. and University of Nebraska-Lincoln Cooperative Extension (2002) *Development of Autonomy in Adolescence*. Lincoln, NE: Institute of Agriculture and Natural Resources, University of Nebraska-Lincoln.

22 Vinnuesa, R. *et al.* (2019) 'The role of artificial intelligence in achieving the sustainable development goals'. arXiv, Cornell University. https://arxiv.org/ftp/arxiv/papers/1905/1905.00501.pdf.

23 Mitchell, T.M. (2018) *Machine Learning*, International edn. New York: McGraw Hill Education.

24 Coole, D. (2017) 'Agency: political science'. Brittanica. https://www.britannica.com/topic/agency-political-theory.

25 Wikipedia Contributors (2019) 'Agent-based model'. Wikipedia. https://en.wikipedia.org/wiki/Agent-based_model.

26 Wikipedia Contributors (2019) 'Subsumption architecture'. Wikipedia. https://en.wikipedia.org/wiki/Subsumption_architecture.

27 Murphy, R. (2019) *Introduction to AI Robotics*. Cambridge, MA: MIT Press.

28 Rao, A.S. and Verweij, G. (eds) (2017) 'Sizing the prize: what's the real value of AI for your business and how can you capitalise?' PwC Global. https://www.pwc.com/gx/en/issues/data-and-analytics/publications/artificial-intelligence-study.html#:~:text=Total%20economic%20impact%20of%20AI%20in%20the%20period%20to%202030&text=AI%20could%20contribute%20up%20to,come%20from%20consumption%2Dside%20effects.

29 Lu, D. (2019) 'Creating an AI can be five times worse for the planet than a car'. *New Scientist*, June. https://www.newscientist.com/article/2205779-creating-an-ai-can-be-five-times-worse-for-the-planet-than-a-car/#:~:text=Training%20artificial%20intelligence%20is%20an,emissions%20of%20an%20average%20car.

30 Daugherty, P.R. and Wilson, H.J. (2018) *Human + Machine: Reimagining Work in the Age of AI*. Brighton, MA: Harvard Business Review Press.

31 Chalmers, D.J. (1996) *The Conscious Mind: In Search of a Theory of Conscious Experience*. New York and Oxford: Oxford University Press.
    Wikipedia Contributors (2019) 'Hard problem of consciousness'. Wikipedia. https://en.wikipedia.org/wiki/Hard_problem_of_consciousness.

32 Penrose, R., Hameroff, S.R., Kak, S. and Tao, L. (eds) (2011) *Consciousness and the Universe: Quantum Physics, Evolution, Brain & Mind*. Cambridge, MA: Cosmology Science Publishers.

33 Searle, J.R. (2002) *Consciousness and Language*. Cambridge: Cambridge University Press (p. 16).

34 Tegmark, M. (2018) *Life 3.0: Being Human in the Age of Artificial Intelligence*. London: Penguin Books (p. 249).

35 merriam-webster.com (2019) 'Definition of automation'. Merriam-Webster. https://www.merriam-webster.com/dictionary/automation.

36 epsrc.ukri.org (n.d.) 'Principles of robotics'. Engineering and Physical Sciences Research Council. https://epsrc.ukri.org/research/ourportfolio/themes/engineering/activities/principlesofrobotics/.

37 Price, A. (2018) 'First international standards committee for entire AI ecosystem: industry recognizes standardization will be essential to broad adoption of AI'. IEC e-tech, 15 March. https://etech.iec.ch/issue/2018-03/first-international-standards-committee-for-entire-ai-ecosystem.

38 Wikipedia Contributors (2020) 'Subject-matter expert'. Wikipedia. https://en.wikipedia.org/wiki/Subject-matter_expert#Domain_expert_(software).

39 Wikipedia Contributors (2020) 'Optical character recognition'. Wikipedia. https://en.wikipedia.org/wiki/Optical_character_recognition.

40 EARTO (2014) 'The TRL scale as a R&I policy tool: EARTO recommendations'. European Association of Research and Technology Organisations. https://www.earto.eu/wp-content/uploads/The_TRL_Scale_as_a_R_I_Policy_Tool_-_EARTO_Recommendations_-_Final.pdf.

41 Strang, G. (2019) *Linear Algebra and Learning from Data*. Wellesley, MA: Wellesey-Cambridge Press.

42 Graham, R.L., Knuth, D.E. and Patashnik, O. (1994) *Concrete Mathematics*, 2nd edn. Boston, MA: Addison Wesley.

43 Strang, G. (2016) *Introduction to Linear Algebra*. Course available via MIT OpenCourseware, Wellesey-Cambridge Press. https://ocw.mit.edu/courses/mathematics/18-06sc-linear-algebra-fall-2011/syllabus/.

44 Knuth, D.E. (2011) *The Art of Computer Programming*, vols 1–4a, Revised edn. Boston, MA: Addison Wesley.

45 https://en.wikipedia.org/wiki/Gaussian_elimination

46 The Editors of Encyclopaedia Britannica (2019) 'Dorothy Vaughan: American mathematician'. Britannica. https://www.britannica.com/biography/Dorothy-Vaughan.

REFERENCES

47. Routledge, R. (2005) 'Bayes's theorem: probability'. Brittanica. https://www.britannica.com/topic/Bayess-theorem.

48. Lee Cooke, R. (2020) 'Andrey Nikolayevich Kolmogorov: Russian mathematician'. Britannica. https://www.britannica.com/biography/Andrey-Nikolayevich-Kolmogorov.

49. Duignan, B. (2010) 'Venn diagram: logic and mathematics'. Britannica. https://www.britannica.com/topic/Venn-diagram.

50. Encyclopaedia Britannica (n.d.) 'Law of total probability: probability'. Britannica. https://www.britannica.com/topic/law-of-total-probability.

51. Wikipedia Contributors (2019) 'Bayesian network'. Wikipedia. https://en.wikipedia.org/wiki/Bayesian_network.

52. Wikipedia Contributors (2019) 'Statistical learning theory'. Wikipedia. https://en.wikipedia.org/wiki/Statistical_learning_theory.

53. The Editors of Encyclopaedia Britannica (2009) 'Probability density function: mathematics'. Britannica. https://www.britannica.com/science/density-function.

54. Bronshteĭn, I.N., Semendiaev, K.A., Musiol, G. and Mühlig, H. (2015) *Handbook of Mathematics*. Heidelberg and New York: Springer.

55. Teukolsky, S.A., Vetterling, W.T., Press, W.H. and Flannery, B.P. (1993) *Numerical Recipes in Fortran*, 2nd edn. Cambridge: Cambridge University Press.

56. Géron, A. (2017) *Hands-On Machine Learning with Scikit-Learn and TensorFlow: Concepts, Tools, and Techniques to Build Intelligent Systems*. Sebastopol, CA: O'Reilly Media.

57. Encyclopaedia Britannica (n.d.) 'Big Data: mining for nuggets of information'. Brittanica. https://www.britannica.com/technology/big-data.

58. Python.org (2019) 'Homepage'. Python. https://www.python.org.

59. Scikit-learn.org (2019) 'Scikit-learn: machine learning in Python – scikit-learn 0.20.3 documentation'. Scikit-learn. https://scikit-learn.org/stable/.
TensorFlow (2019) 'Homepage'. TensorFlow. https://www.tensorflow.org.

60. McCulloch, W.S. and Pitts, W. (1943) 'A logical calculus of the ideas immanent in nervous activity'. *Bull. Math. Biophys.*, 5 (4), 115–133.

61. Pitts, W. and McCulloch, W.S. (1947) 'How we know universals the perception of auditory and visual forms'. *Bull. Math. Biophys.*, 9 (3), 127–147.

62. Minsky, M.L. and Papert, S.A. (1988) *Perceptrons: An Introduction to Computational Geometry*, Extended edn. Cambridge, MA: MIT Press.

63. Rosenblatt, F. (1962) *Principles of Neurodynamics: Perceptrons and the Theory of Brain Mechanisms*. Washington, DC: Spartan Books.

64. OpenAI (2019) 'Solving Rubik's Cube with a robot hand'. OpenAI. https://openai.com/blog/solving-rubiks-cube/.

65. Hunt, E. (2016) 'Tay, Microsoft's AI chatbot, gets a crash course in racism from Twitter'. *The Guardian*, 24 March. https://www.theguardian.com/technology/2016/mar/24/tay-microsofts-ai-chatbot-gets-a-crash-course-in-racism-from-twitter.

66 UCL (2019) 'Health and social care'. AI for People and Planet, University College London. https://www.ucl.ac.uk/artificial-intelligence/our-research/health-and-social-care.

67 Griffiths, S. (2016) 'This AI software can tell if you're at risk from cancer before symptoms appear'. *Wired*. www.wired.co.uk/article/cancer-risk-ai-mammograms.

68 Matheson, R. (2019) 'Study: social robots can benefit hospitalized children'. *MIT News*, 26 June. https://news.mit.edu/2019/social-robots-benefit-sick-children-0626.

69 Hodson, H. (2016) 'Google's DeepMind AI can lip-read TV shows better than a pro'. *New Scientist*. https://www.newscientist.com/article/2113299-googles-deepmind-ai-can-lip-read-tv-shows-better-than-a-pro/#ixzz6SOItlk00.

70 Niestadt, M., Debyser, A., Scordamaglia, D. and Pape, M. (2019) *Artificial Intelligence in Transport: Current and Future Developments, Opportunities and Challenges*. Briefing. European Parliamentary Research Service. https://www.europarl.europa.eu/RegData/etudes/BRIE/2019/635609/EPRS_BRI(2019)635609_EN.pdf.

71 Association for Safe International Road Travel (2018) 'Road safety facts'. Association for Safe International Road Travel. https://www.asirt.org/safe-travel/road-safety-facts/.

72 Ozay, N. and Panagou, D. (n.d.) 'Robot teams & swarms'. M Robotics, University of Michigan Robotics Institute. https://robotics.umich.edu/research/focus-areas/robot-teams-swarms/.

73 Toole, J. (2019) 'Combining artificial intelligence and augmented reality in mobile apps'. Heartbeat. https://heartbeat.fritz.ai/combining-artificial-intelligence-and-augmented-reality-in-mobile-apps-e0e0ad2cfddc.

74 Condon, S. (2018) 'IBM researchers build AI-powered prototype to help small farmers test soil'. ZDNet. https://www.zdnet.com/article/ibm-researchers-help-small-farmers-test-soil-with-ai/.

75 Villani, C. (2018) *For a Meaningful Artificial Intelligence: Towards a French and European Strategy*. https://www.aiforhumanity.fr/pdfs/MissionVillani_Report_ENG-VF.pdf.

76 Congressional Research Service (2019) 'Artificial intelligence and national security'. Federation of American Scientists. https://fas.org/sgp/crs/natsec/R45178.pdf.

77 Fryer-Biggs, Z. (2019) 'Coming soon to a battlefield: robots that can kill'. *The Atlantic*, 3 September. https://www.theatlantic.com/technology/archive/2019/09/killer-robots-and-new-era-machine-driven-warfare/597130/.

78 Katwala, A. (2020) 'The US Air Force is turning old F-16s into pilotless AI-powered fighters'. *Wired*, 27 June. https://www.wired.co.uk/article/f-16-us-air-force-qf-16.

79 US DOD (2020) *DOD Dictionary of Military and Associated Terms*. www.jcs.mil/Portals/36/Documents/Doctrine/pubs/dictionary.pdf.

80 Magoulas, R. and Swoyer, S. (2020) 'AI adoption in the enterprise 2020'. *O'Reilly*, 18 March. https://www.oreilly.com/radar/ai-adoption-in-the-enterprise-2020/.

81. Wikipedia Contributors (2019) 'Digital twin'. Wikipedia. https://en.wikipedia.org/wiki/Digital_twin.
82. ECMWF (n.d.) 'Homepage'. European Centre for Medium Range Weather Forecasts. https://www.ecmwf.int.
83. Met Office (2019) 'Homepage'. Met Office. https://www.metoffice.gov.uk.
84. Batchelor, G.K. (1969) 'Computation of the energy spectrum in homogeneous two-dimensional turbulence'. *Phys. Fluids*, 12 (Suppl. II), 233–239.

FURTHER READING

MACHINE LEARNING

Chalmers, D. (1996) *The Conscious Mind in Search of a Fundamental Theory*. New York: Oxford University Press.

Daugherty, P.R. and Wilson, H.J. (2018) *Human + Machine: Reimagining Work in the Age of AI*. Boston, MA: Harvard Business Review Press.

Frankish, K. and Ramsey, W. (2014) *The Cambridge Handbook of Artificial Intelligence*. Cambridge: Cambridge University Press.

Géron, A. (2019) *Hands-On Machine Learning with Scikit-Learn, Keras and TensorFlow: Concepts, Tools, and Techniques to Build Intelligent Systems*, 2nd edn. Sebastopol, CA: O'Reilly Media.

Ghallab, M., Nau, D. and Traverso, P. (2004) *Automated Planning: Theory and Practice*. San Francisco, CA: Elsevier.

Kurzweil, R. (2005) *The Singularity is Near: When Humans Transcend Biology*. London: Duckworth Overlook.

Mitchell, T.M. (1997) *Machine Learning*, International edn. Singapore: McGraw Hill Education.

Murphy, K.P. (2012) *Machine Learning: A Probabilistic Perspective*. Cambridge, MA: MIT Press.

Rouhiainen, L. (2018) *Artificial Intelligence: 101 Things You Must Know Today About Our Future*. CreateSpace Independent Publishing Platform.

Russell, S. and Norvig, P. (2016) *Artificial Intelligence: A Modern Approach*, Global edn. Harlow: Pearson Education Ltd.

Schwab, K. (2016) *The Fourth Industrial Revolution*. London: Penguin Random House.

Tegmark, M. (2017) *Life 3.0: Being Human in the Age of Artificial Intelligence*. London: Penguin Books.

Theobald, O. (2017) *Machine Learning for Absolute Beginners: A Plain English Introduction*. Scatterplot Press.

ADVANCED THEORETICAL TEXT

Shalev-Shwartz, S. and Ben-David, S. (2014) *Understanding Machine Learning: From Theory to Algorithms*. New York: Cambridge University Press.

INDEX

abstraction 2, 6, 116
agent-based modelling 26, 29, 113, 124
agent schematic 36, 70, 85, 120
agent types 29–31
    goal-based 30
    learning agent 10, 27, 31, 36, 70, 85
    model-based reflex 30
    reflex 30
    utility-based reflex 30–1
AGI (artificial general intelligence) 3, 4, 8, 14, 41, 42, 44
Agile approach 45, 46, 67, 99–101, 116
AI agent 24–5, 26–31, 36, 80, 82, 86
AI algorithm 44, 66, 89, 90–3, 98, 115
AI case studies 119–33
    Cambridge spin-out start-up 130–3
    warehouse performance 119–24
    weather 124–30
AI ethics 8, 14–18, 20, 21, 22, 44, 95, 100
AI in use in industry 102–18
    agriculture 114–15
    education 105–6
    engineering 112–14
    finance 105
    health and social care 103–5
    logistics 106–7
    media and entertainment 108–10
    military 115–16
    research and development 102–3
    retail 107–8
    sales and marketing 116–17
    transportation 110–12
AI project team 45, 46, 99–101
algorithms 88–95
    AI/ML algorithms 90–3
    deep learning algorithms 93–4
    definition and types 89–90
    ethical use 94–5
    machine learning and artificial intelligence algorithms 90–3
    self-learning algorithms 90
AlphaGo 8, 85
Artificial Intelligence: A Modern Approach 3
artificial intelligence and robotics 26–36
    AI intelligent agent 26–31
    what is a robot? 31–6
    what is an intelligent robot? 36
Asilomar Principles 17–18
augmenting humans 96–7
augmenting machines 97

backpropagation 93
Batchelor, George 125, 127, 128
benefits, challenges and risks of AI 37–48
    benefits 39–43
    challenges 43–4
    funding AI projects 47
    opportunities 46–7
    risks 44–6
    sustainability 37–9

cloud computing 14, 41, 43, 46, 86–7
CNN (Convolutional Neural Network) 93
Convenience for You is Independence for Me 105
COVID-19 95, 98
creativity 2, 22, 37, 40, 110

data preparation 66–9
data scientist 45, 77, 97, 98
Daugherty, Paul R. and Wilson, H. James: *Human + Machine: Reimagining Work in the Age of AI* 40
DBN (deep belief network) 94
decision tree 41, 75, 88, 90, 91
Deep Blue 7, 31
diffusion of innovation 9
digital twin 28, 29, 36, 56, 66, 69, 113, 124
domain experts 46, 65, 120, 122–3, 125, 127, 128, 130, 133

Engineering and Physical Sciences Research Council 15, 44
ethical and sustainable human and artificial intelligence 1–25
    definition of artificial intelligence 3–4
    definition of human intelligence 1–3
    history of artificial intelligence 4
    machine learning 24–5
    sustainable AI 23–4
Ethics Guidelines for Trustworthy AI 20

'fit for purpose' 45, 46, 67, 76, 94, 120, 121

free will 19, 27
Future of Life Institute 17–18

GAN (Generative adversarial network) 94
goal-based agent 30
Graham, R.L., Knuth D.E., and Patashnik, O.: *Concrete Mathematics* 54

how to build a machine learning toolbox 49–87
  agents' functionality 85–6
  learning from data 49–80
  neural networks 80–5
  using the cloud 86–7
human-centric ethical purpose 17, 37, 38, 42, 96, 97
human learning 3, 121, 122, 123, 128
human-only systems 40
human plus machine systems 42
hyper-parameters 75

IA (intelligent automation/augmentation) 96–7
industrial revolutions 8–9, 10, 46, 96, 102–17
intelligent agent 7, 10, 26–31, 36, 107
intergenerational equity 37–8

K-means 90, 92
k-NN (k-Nearest Neighbour) 41, 90, 91–2
Knuth, D. E.: *The Art of Computer Programming* 54
Kurzweil, Ray 4, 6, 46, 47
  The Singularity is Near 4, 46, 99

learning agent 10, 27, 31, 36, 70, 85
learning from data 10, 25, 49–80, 120
Lighthill, Sir James 6, 42
linear algebra 50–4
linear regression 90–1
logic 4, 6, 90, 106

logistic regression 91
LSTM (long short-term memory) 94

machine-only systems 40–1
management, roles and responsibilities 96–101
  future directions 99
  humans and machines working together 96–9
  'Learning from experience' Agile approach 99–101
Mitchell, Tom 6, 24, 50
  Machine Learning 88
ML (machine learning) 3, 6, 10, 14, 24–5, 31, 49–88, 98, 121–2, 125
ML algorithm 70–3, 88, 90–3
MLP NN (Multilayer Perceptron neural network) 93
model-based reflex agent 30

Naive Bayes 90, 92
narrow AI 41, 82, 88, 130
NASA 47, 55, 127
NN (neural network) 5, 6, 25, 36, 70, 80–5, 93

open-source 46, 52, 53, 78, 80, 86
Optalysys 130–3
over-fitting 75–7, 91

Perceptron 82, 84, 93
probability and statistics 5, 50, 57–66
Python 43, 66, 77, 78

random forest 91
RBM (Restricted Boltzmann machine) 94
recommendation engines 90
reflex agent 30
RNN (Recurrent Neural Network) 93
robotic paradigms 32–6
  hierarchical 34
  hybrid 35–6
  reactive 34–5

Russell, Stuart 42
  Human Compatible: Artificial Intelligence and the Problem of Control 24
Russell, Stuart and Norvig, Peter 5, 10, 26, 27, 28, 36, 130
  Artificial Intelligence: A Modern Approach 3, 50

Samuel, Arthur Lee 6
scientific method 4, 10, 28, 41, 44, 88, 94, 102, 121–2, 127, 129, 133
SFIA (Skills for the Information Age) Foundation framework 101
Sternburg, Robert 2
Strang, Gilbert 50, 54
strong AI 4, 41, 42, 44
sustainability 20, 23–4, 37–9, 119, 124, 130
  design 39
  sustainable AI 23
  three pillars 38
  United Nations sustainability goals 17, 44
SVM (support vector machine) 41, 70, 90, 92

Tegmark, Max 42
  Life 3.0 99
The Physics of Fluids 127
TRL (Technology Readiness Level) 47, 55, 127, 133
trustworthy AI 14, 16, 17, 18, 20, 21, 22, 23
Turing test 4
Turing, Alan 4, 5

under-fitting 75–7
universal design 9, 19, 20
utility-based reflex agent 30–1

vector calculus 50, 54–6, 57, 66, 78, 85, 128
visualising data 69, 76–80, 128

Printed in the USA
CPSIA information can be obtained
at www.ICGtesting.com
LVHW081650210724
786107LV00007B/560

9 781780 175287